The Principles of
PETROLOGY

The Principles of
PETROLOGY

An Introduction to the Science of Rocks

by

G. W. TYRRELL
A.R.C.Sc., F.G.S., F.R.S.E., D.Sc. (Glasgow)

*Lecturer in Geology in the
University of Glasgow*

LONDON
CHAPMAN AND HALL

First published 1926
by Methuen & Co Ltd
Reprinted twenty-one times
Reprinted as a Science Paperback 1978
Chapman & Hall Ltd
11 New Fetter Lane, London EC4P 4EE
Reprinted 1979

Redwood Burn Limited
Trowbridge & Esher

ISBN 0 412 21500 4

ISBN-13: 978-0-412-21500-1   e-ISBN-13: 978-94-011-6026-1
DOI: 10.1007/978-94-011-6026-1

Distributed in the USA by
HALSTED PRESS
A DIVISION OF JOHN WILEY & SONS INC.
NEW YORK

PREFACE

IN this book the task of summarising modern petrology from the genetic standpoint has been attempted. The scale of the work is small as compared with the magnitude of its subject, but it is nevertheless believed that the field has been reasonably covered. In conformity with the genetic viewpoint petrology, as contrasted with petrography, has been emphasised throughout; and purely descriptive mineralogical and petrographical detail has been omitted.

Every petrologist who reads this book will recognise the author's indebtedness to Dr. A. Harker and Dr. A. Holmes, among British workers; to Prof. R. A. Daly, Dr. H. S. Washington, and Dr. N. L. Bowen, among American petrologists; and to Prof. J. H. L. Vogt, Prof. V. M. Goldschmidt, Prof. A. Lacroix, and Prof. P. Niggli, among European investigators. The emphasis laid on modern views, and the relative poverty of references to the works of the older generation of petrologists, does not imply any disrespect of the latter. It is due to recognition of the desirability of affording the petrological student a newer and wider range of reading references than is usually supplied in this class of work; for references tend to become stereotyped as well as text and illustrations. Furthermore it is believed that all that is good and living in the older work has been incorporated, consciously or unconsciously, in the newer. The materials provided by the older investigators have now become the corner stones of the modern structure of petrological science, and do not need to be specifically pointed out in every new textbook.

Many of the illustrations are new. Others have been redrawn and modified from publications to which full acknowledgment is made in the text.

It is hoped that the book will appeal to students of geology who have acquired an elementary knowledge of the science, and who are going on to more advanced work; to advanced students and post-graduate researchers; and to workers in other branches of geology who desire a conspectus of the present state of the science of petrology. Furthermore it is believed that the book will prove useful to workers engaged in technical and applied aspects of geology, in all of which some knowledge of rocks in general is fundamental.

To Prof. J. W. Gregory, D.Sc., F.R.S., my cordial thanks are due for reading the MS. and, in his constructive criticism, making many valuable suggestions and corrections which have been incorporated in the book.   I am also indebted to Dr. M. A. Peacock, Carnegie Research Fellow in the University of Glasgow, who kindly read the manuscript keeping in mind the student's point of view, and thereby contributed materially to any clarity it may possess.

G. W. TYRRELL

UNIVERSITY OF GLASGOW

September, 1926

PREFACE TO SECOND EDITION

THE opportunity afforded by a second edition has been taken to correct a number of minor errors, to revise as far as is possible within the frame of the present work, and to insert a few additional references.   The author is indebted to those reviewers and correspondents who have kindly pointed out errors and omissions.

G. W. T.

UNIVERSITY OF GLASGOW
December, 1929

CONTENTS

vii

PART II

THE SECONDARY ROCKS

CHAPTER X

THE RESIDUAL DEPOSITS

CHAPTER XI

SEDIMENTARY ROCKS : MINERALOGICAL, TEXTURAL, AND STRUCTURAL CHARACTERS

CHAPTER XII

SEDIMENTARY ROCKS. DESCRIPTIVE

CHAPTER XIII

DEPOSITS OF CHEMICAL ORIGIN

CHAPTER XIV

DEPOSITS OF ORGANIC ORIGIN

PART III

THE METAMORPHIC ROCKS

CHAPTER XV

METAMORPHISM

LIST OF DIAGRAMS

xi

THE PRINCIPLES OF PETROLOGY

CHAPTER I

INTRODUCTION

THE SCIENCE OF PETROLOGY—Petrology is the science of
rocks, that is, of the more or less definite units of which the
earth is built. In the nature of things the study is limited
to the materials of the accessible crust, although we have in
meteorites samples of rocks which must be identical with, or
analogous to, those composing the interior of the earth.
The science deals with the modes of occurrence and origin
of rocks, and their relations to geological processes and history. Petrology is thus a fundamental part of geological
science, dealing, as it does, with the materials the history of
which it is the task of geology to decipher.

Rocks can be studied in two ways : as units of the earth's
crust, and therefore as documents of earth history, or as
specimens with intrinsic characters. The study of rocks as
specimens may properly be designated *petrography*. *Petrology* is, however, a broader term which connotes the philosophical side of the study of rocks, and includes both
petrography and *petrogenesis*, the study of origins. Petrography comprises the purely descriptive part of the science
from the chemical, mineralogical, and textural points of view.
As we must gain an exact knowledge of the units we have
to deal with before we can study their broader relations to
geological processes and their origins, petrography is a
necessary pre-requisite to petrology, and must be carried
on as far as possible by quantitative methods, as in other
physical and chemical sciences.

The term *lithology* is nearly synonymous with petrology,

I

but is now seldom used. Etymologically, it means the science of stones; and, accordingly, there is a tendency, which should be encouraged, to use the term to indicate the study of stones in engineering, architecture, and in other fields of applied geology. It would be equally appropriate and correct to speak of the lithology of conglomerates and breccias, when dealing with the stones contained in these rocks.

Broadly speaking, petrology is the application of the principles of physical chemistry to the study of naturally-occurring earth materials, and may therefore be regarded as the natural history branch of physical chemistry. Viewed thus, petrology is again seen as a subdivision of geology. The too exclusive study of petrography sometimes tends to obscure this relationship, and hides the fact that petrology is intimately connected with a host of fascinating geological problems.

THE EARTH ZONES—Regarded as a whole, the earth is a sphere of unknown material surrounded by a number of thin envelopes. The inaccessible heavy interior is known as the *barysphere*. This is followed outwardly by the *lithosphere*, the thin, rocky crust of the earth; then by a more or less continuous skin of water, the *hydrosphere;* and finally by the outermost envelope of gas and vapour, the *atmosphere*. Other zones have been distinguished and named for special purposes. The zone of igneous activity and lava formation, situated between the lithosphere and the barysphere, is the *pyrosphere;* and Walther has distinguished the living envelope which permeates the outermost zones, as the *biosphere*. A zone towards the base of the lithosphere which can sustain little or no stress has been called the *asthenosphere* (sphere of weakness) by Barrell; and the zone in which crustal movements originate has been named the *tectonosphere* by certain Continental geologists.

THE BARYSPHERE—While the interior of the earth, of course, is inaccessible to direct observation, many facts have been indirectly ascertained about its structure and composition. It must, for example, be hot. Observations in wells, mines, and borings, show that there is a downward increase of temperature, which is variable in different parts of the earth, and averages 1° C. for every 31·7 metres of depth, or roughly 50° C. per mile, in Europe. In North America the temperature gradient is much gentler, only 1° C. to 41·8 metres of depth, or 38° C. per mile. If these gradients continue indefinitely, enormous temperatures must prevail in the interior of the earth, but it is probable that the rate of increase falls off with depth.[1] Similarly, very great pressures must occur, even at moderate depths within the crust.

The barysphere must also be composed of heavy materials. The density of the earth as a whole is 5·6, but the average density of known rocks of the lithosphere is only 2·7. Hence the average density of the barysphere must be somewhat greater than 5·6.

Several considerations serve to show that the barysphere must be rigid, with a rigidity greater than that of the finest tool-steel. In the early days of geology it was believed that the thin solid crust rested or floated on a molten interior; but when attention was given to geophysical matters it was soon shown that, under these circumstances, the thin crust would experience great distortion in response to the attraction of the moon, and, furthermore, owing to internal friction, rotation could not long be maintained.

A strong confirmation of the rigidity of the barysphere is afforded by the study of earthquake vibrations. A heavy shock, say, in New Zealand, is recorded by seismometers in Britain about 21 minutes later, the vibrations having travelled by a more or less direct path through the barysphere. This speed of wave propagation is consistent only with high rigidity in most of the interior of the earth.

COMPOSITION OF THE EARTH SHELLS—The earth has been called a projectile of nickel-steel covered with a slaggy crust.[2] It is probable that all the planets and planetoids of the solar system have essentially the same composition; hence the wandering fragments of planetary matter known as *meteorites*, or shooting stars, which the earth sweeps up as it revolves in its orbit, are of particular interest in connection with the present topic. Meteorites fall upon the earth's surface in masses which vary in size from the finest dust to huge blocks weighing many tons.

Meteorites are divided into three main groups which pass gradually one into the other :—

[1] R. A. Daly, *Amer. Journ. Sci.* (5), vol. v, 1923, p. 352.
[2] J. W. Gregory, *Report Brit. Assoc.*, Leicester, 1907, p. 494.

1. *Siderites.*—The iron meteorites, consisting almost entirely of iron alloyed with nickel.

2. *Siderolites.*—Mixtures of nickel-iron and heavy basic silicates, such as olivine and pyroxene.

3. *Aerolites.*—The stony meteorites, consisting almost entirely of heavy basic silicates, olivine and pyroxenes, and resembling some of the rarer and most basic types of terrestrial igneous rocks. There are small amounts of sulphur, phosphorus, carbon, and other elements in meteorites, which, however, may be disregarded in the present connection.

Professor J. W. Gregory [1] shows that if all known meteorites are considered, the iron group far outweighs the stony group. The stony meteorites fall in greater abundance, but the siderites fall in such large masses that they bulk much greater than the aggregates of small aerolites. Hence the relative masses of the different types of meteorites support the above-cited view of the composition of the earth.

From geophysical data based on the distribution of density, earthquake vibrations, etc., Williamson and Adams [2] have arrived at the conception of earth composition illustrated by Fig. 1 A. In this the earth is shown to be built of four layers : (1) a thin surface crust of light silicates and silica ; (2) a zone of heavy silicates (peridotite) which, of density 3·3 in its uppermost layers, is of density 4·35 in its lowest part at a depth of 1600 kms. ; (3) increasing admixture of nickel-iron leads to a zone consisting of material similar to siderolites (pallasite) [3] which, with a rapidly-mounting proportion of nickel-iron, passes into ; (4) the purely metallic core. The actual composition of meteorites, as given above, supports this hypothesis. According to this view the earth is conceived as the result of a gigantic metallurgical operation analogous to the smelting of iron, with the more or less complete separation of metal and silicate slag.

Professor V. M. Goldschmidt [4] has put forward a view of earth composition which differs from the above chiefly in the intercalation of a zone of metallic sulphides and oxides

[1] *Op. cit. supra.*

[2] *Proc. Washington Acad. Sci.*, vol. xiii, 1923, pp. 413-28.

[3] *Pallasite*, siderolites containing olivine crystals in a network of nickel-iron.

[4] *Vidensk. Selsk. Skr. Kristiania*, 1922, No. 11.

(Fig. 1 B) between the nickel-iron core and the shell of heavy compressed silicates. In this conception the analogy of copper smelting is kept in mind, involving the separation of metal, sulphide-matte, and slag.

CHEMICAL COMPOSITION OF THE CRUST—The outer crust of the earth down to a depth of 10 miles or so, consists of igneous rocks and metamorphic rocks, with a thin, interrupted mantle of sedimentary rocks resting on them. According to Clarke and Washington [1] the lithosphere down to a

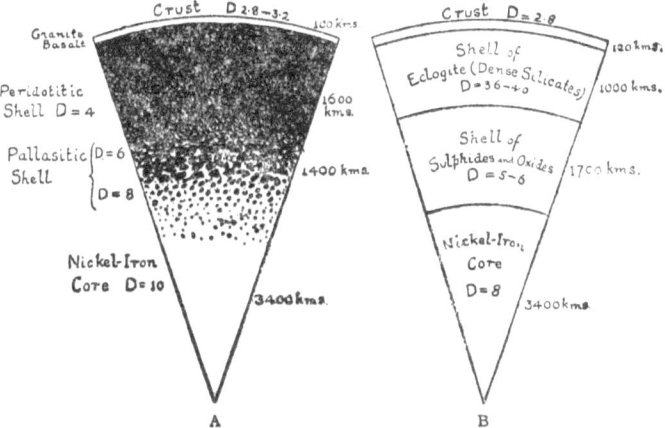

FIG. 1.—SECTIONS THROUGH THE EARTH.
(A) According to Williamson and Adams.
(B) According to Goldschmidt (text, p. 4).

depth of 10 miles is made up of igneous rocks, 95 per cent. ; shale, 4 per cent. ; sandstone, 0·75 per cent. ; and limestone, 0·25 per cent. The metamorphic rocks, which are derived from igneous and sedimentary rocks by alteration under heat and pressure, are regarded as belonging to their initial types, and are neglected in the following calculation. The average chemical composition of each class of rock has been obtained by computation from a large number of existing analyses ; and if these are combined in the proportions given above the

[1] *Prof. Paper* 127, *U.S. Geol. Surv.*, 1924, p. 32.

following figures are obtained (col. 1), representing the composition of the lithosphere down to the 10-mile limit of depth. Column 2 shows the composition as recast into the form of oxides:—

TABLE I

1			2		
Oxygen	.	46·71	SiO_2	.	59·07
Silicon	.	27·69	Al_2O_3	.	15·22
Aluminium	.	8·07	Fe_2O_3	.	3·10
Iron	.	5·05	FeO	.	3·71
Calcium	.	3·65	MgO	.	3·45
Sodium	.	2·75	CaO	.	5·10
Potassium	.	2·58	Na_2O	.	3·71
Magnesium	.	2·08	K_2O	.	3·11
Titanium	.	·62	H_2O	.	1·30
Hydrogen	.	·14	CO_2	.	·35
Phosphorus	.	·13	TiO_2	.	1·03
Carbon	.	·094	P_2O_5	.	·30
Manganese	.	·09	MnO	.	·11
Sulphur	.	·052	Rest [1]	.	·44
Barium	.	·050			
Remaining elements		·244			100·00
		100·00			

On the assumption that the 10-mile crust is composed entirely of igneous rocks, the results of the computations only differ, on the whole, from those given above, in the second figure of decimals.

It thus appears that fifteen elements between them make up 99·75 per cent. of the earth's crust, and that the majority of the elements which are important in human affairs are included in the remainder, being present in the crust in amounts of the order of 1/100th of 1 per cent.

It has been asserted that the Clarke-Washington method seriously over-estimates the acidity (amount of silica) of the crust, since no allowance is made for the relative amounts of the different kinds of igneous rock composing the average. Rocks of acid composition are largely restricted to the continents, and are probably underlain at shallow depths by

[1] " Rest " includes ZrO_2, ·04; Cl, ·05; F, ·03; S, ·06; $(Ce, Y)_2O_3$, ·02; Cr_2O_3, ·05; V_2O_3, ·05; NiO, ·03; BaO, ·05; SrO, ·02; Li_2O, ·01; Cu, ·01; C, ·04.

basic rock (basalt, etc.). The floors of the oceans also probably consist of basalt. Hence, perhaps, the average crustal
rock should be regarded as considerably more basic than is
allowed by the Clarke-Washington method.[1]

ROCKS AND THEIR COMPOSITION.—Rocks have already
been defined as the more or less definite units constituting
the earth's crust, but in popular usage the term *rock* denotes
any hard, solid material derived from the earth. In geology,
however, the term is often used without reference to the
hardness, or state of cohesion, of the material; and sand,
clay, or peat, are thus just as much rocks in the scientific
sense as granite and limestone.[2]

When solid rocks are examined closely they are found to
consist largely of fragments of simpler chemical composition,
which are called *minerals*, or strictly, *mineral species*. Hence
we arrive at another definition of rocks as aggregates of
mineral particles. It is important to distinguish the two
uses of the term mineral. It is used in a perfectly legitimate
sense by the ordinary man, the miner, prospector, quarry-
master, lawyer, landowner, and the scientific mineralogist,
to indicate materials, such as coal, slate, clay, etc., which
are won from the earth's crust, but which, in the strict
petrological sense, are rocks. We speak also of " mineral
waters " and " mineral oil " in the same way. Natural
glasses and other amorphous materials which occur as rocks
and in rocks may also be regarded as minerals in this sense.

The materials of which rocks are largely composed are,
however, *mineral species*, i.e. natural inorganic substances
with a definite chemical composition, or a definite range of
chemical composition, and a regular internal molecular structure, which manifests itself under favourable circumstances
by the assumption of regular crystalline form, and the
possession of definite optical and other measurable properties.

THE ROCK-FORMING MINERALS — The table on page 6,
column 2, shows the principal chemical elements present in
the earth's crust combined with oxygen, the most abundant

---

[1] T. Crook, *Nature*, 19 Aug., 1922, pp. 253-5.
[2] This usage, however, does violence to the accepted meaning of a
popular term. See Prof J. W. Gregory, " The Scientific Misappropria-
tion of Popular Terms," *Rept. Brit. Assoc. Adv. Sci. Portsmouth*, 1911,
pp. 742-7.

element, to form oxides; and this, by the way, is the most convenient and usual method of presenting the chemical analyses of rocks. These elements, however, fall into various combinations from which the rock-forming minerals arise, in which oxides, as such, are only of secondary importance. Silicates are the most abundant compounds constituting the rock-forming minerals; oxides come next; then carbonates, phosphates, sulphates, etc., in greatly diminished importance. Among the elements listed on page 6, only iron (in the basalt of Disko, Greenland), carbon (as diamond and graphite), and sulphur (volcanic action; decomposition of sulphates and sulphides), occur native.

It is probable that 99·9 per cent. of the earth's crust is composed of only about twenty minerals out of the thousand or so which have been described. These are the rock-forming minerals *par excellence*. Referring first to the silicates, the *felspars* are by far the most abundant and important group, not only of the silicates, but of the rock-forming minerals in general. The chief members are *orthoclase* and *microcline*, both silicate of potassium and aluminium; and the various *plagioclases*, which are mixtures in all proportions of the two end-members, *albite*, silicate of sodium and aluminium, and *anorthite*, silicate of calcium and aluminium. Allied to the felspars, but containing less silica in proportion to the bases present, are the *felspathoid* minerals, of which the most important are *nepheline*, silicate of sodium and aluminium (corresponding to albite among the felspars), and *leucite*, silicate of potassium and aluminium (corresponding to orthoclase among the felspars). The mineral *analcite*, a silicate of sodium and aluminium with combined water, and properly belonging to the group of *zeolites*, may nevertheless take its place as a rock-forming mineral with the felspathoids.

The *mica* group forms a link between the alkali-alumina silicates above mentioned and the heavier and darker ferro-magnesian silicates to be described later. Of micas proper, the two chief members are the white mica, *muscovite*, silicate of potassium and aluminium with some hydroxyl, and the dark mica, *biotite*, silicate of potassium, aluminium, magnesium, and iron, with hydroxyl. *Chlorite* is the green hydrated silicate of magnesium and iron, of micaceous affinities, which is the most familiar alteration product of biotite and other ferro-magnesian minerals.

Among the ferro-magnesian minerals proper there are the three main groups of the pyroxenes, amphiboles, and olivines. The *pyroxenes* are metasilicates of calcium, magnesium, and iron, of which the two chief members are the *orthopyroxenes* (*enstatite* and *hypersthene*), simple metasilicates of magnesium and iron ; and *augite*, the monoclinic pyroxene, a complex m :tasilicate of calcium, magnesium, iron, and aluminium. The *amphiboles* form a parallel group to the pyroxenes, but with different crystal habit. The chief member is *hornblende*, a mineral of composition similar to that of augite, but usually richer in calcium. The *olivines* are simple orthosilicates of magnesium and iron, and stand in the same relation to pyroxenes and amphiboles as the felspathoids do to the felspars. *Serpentine* is the hydrated alteration product of olivine and other ferro-magnesian minerals.

Numerous other silicates occur as rock-forming minerals, but it is only necessary here to mention the *garnets*, a varied isomorphous series, chiefly silicates of iron, magnesium, calcium, and aluminium ; *epidote*, silicate of calcium, iron, and aluminium, an abundant alteration product of lime-rich silicate minerals ; *andalusite*, *kyanite*, and *sillimanite*, all simple silicates of aluminium ; *cordierite*, silicate of magnesium, iron, and aluminium ; and *staurolite*, silicate of iron and aluminium. The five last-named minerals are characteristic of the metamorphic group of rocks.

Among the oxide minerals only four need be mentioned as prominent rock-formers. *Quartz*, the dioxide of silicon, is, perhaps, the most abundant mineral next to the felspars. Impure colloidal silica, especially the dark variety known as *chert*, also forms large rock masses. The oxides of iron come next : *magnetite* (Fe_3O_4) is very widely distributed in rocks in small quantities ; *hæmatite* (Fe_2O_3) and *limonite* (Fe_2O_3, nH_2O) form the universal red, brown, and yellow colouring matters of rocks. *Ilmenite*, the oxide of iron and titanium ($FeTi)_2O_3$, is, perhaps, even more widely distributed in rocks than magnetite.

Of carbonates, the minerals *calcite*, the carbonate of calcium, and *dolomite*, the carbonate of calcium and magnesium, are by far the most abundant, and are the chief minerals of the important group of the limestone rocks. One phosphate, the mineral *apatite*, phosphate of calcium with some combined fluorine or chlorine, is universally distributed in small amounts; and one sulphide, the ubiquitous *pyrites*, disulphide of iron, is a common rock-forming mineral. Two sulphates, *gypsum*, sulphate of calcium with combined water, and *barytes*, sulphate of barium, occasionally form rock masses; as also one chloride, common salt or *halite*, the chloride of sodium.

For information regarding the crystallographical, optical, and other properties of the rock-forming minerals the reader is referred to one of the numerous standard textbooks on that subject.

THE CLASSIFICATION OF ROCKS—Whatever theory of earth origin be held it is at least certain that all parts of the original surface of the earth passed through a molten stage, and that the first solid material which existed was derived from a melt or *magma*. This original crust is nowhere exposed on the present surface, but all subsequently-formed rocks, in the first instance, have been produced either from this, or from later irruptions of molten matter. Rocks formed by the consolidation of molten magma are said to be *Primary or Igneous*.

After the solidification of the original crust, and the formation of the hydrosphere and atmosphere, the waters and the air, both probably of much greater chemical potency than now, began to attack the primary rocks. Disintegrative action produced loose debris, and chemical action produced both debris and material in solution. The loose fragments would be swept away by water and wind, and would ultimately collect in the hollows of the crust, where also the waters and soluble matters would be found. The collected debris deposited from suspension in water or air would finally be cemented into hard rock, and would be thus added to the solid crust. Under suitable circumstances the soluble matter likewise would be precipitated, either directly, or indirectly through the agency of organisms, the latter, of course, in somewhat later geological times. The rocks thus produced would eventually become solid and help to build up the crust. These processes have gone on through-

out geological time, the newer increments of the crust undergoing attack as well as the older parts. Hence it may be
that some of the material has gone through many successive
cycles of change.

The rocks formed in these ways are called *Secondary*, because
they are composed of second-hand or derived materials. They
may be divided into *Sedimentary*, *Chemical*, or *Organic*,
according to the process by which they received their most
distinctive characters.

Finally, both Primary and Secondary rocks may be subjected to earth movements which carry them down to depths
in the crust where they are acted upon by great heat and
pressure. By these agencies the rocks are partly or wholly
reconstituted; their original characters are partly or wholly
obliterated, and new ones impressed upon them. Rocks
thus more or less completely changed from their original
condition are known as the *Metamorphic* rocks.

We thus arrive at the time-honoured three-fold classification of rocks according to their modes of origin into Igneous,
Secondary (Sedimentary), and Metamorphic. The Primary
rocks are distinguished by the presence of crystalline minerals
which interlock one with the other, or are set in a minutely-
crystalline paste, or in a glass. They show signs, as do
present-day lavas, of having cooled from a high temperature.
They are usually massive, unstratified, unfossiliferous, and
often occupy veins and fissures breaking across other rocks,
which they have obviously heated, baked, and altered.

The Secondary rocks are composed of clastic and precipitated materials, or of substances of organic character and
origin. The materials are often loose and unconsolidated,
or are welded together by pressure or by a cementing substance into a solid rock. They are further distinguished by
the frequent presence of bedding or stratification, organic
remains (fossils), and other marks indicative of deposition
from water or air in the sea or on land.

The Metamorphic rocks present characters which, in some
respects, are intermediate between those of the Primary and
Secondary rocks. Great heat and pressure cause recrystallisation; hence, like the Primary rocks, they often consist of
interlocking crystals. Furthermore, pressure causes the development of more or less regular layers, folia, or banding,

in which the Metamorphic rocks resemble those of Secondary origin. Since the Metamorphic rocks are formed from pre-existing igneous or sedimentary rocks they often retain traces of their original structures.

Mr. T. Crook has recently formulated a genetic classification, in which the rocks are arranged according to a geological grouping of processes. He divides rocks into two great classes[1]: 1. *Endogenetic*, formed by processes of internal origin which operate deepseatedly or from within outwards (with respect to the crust). High temperature effects are predominant, and the water associated with the processes is partly of magmatic origin. 2. *Exogenetic*, formed by processes of external origin, operating superficially or from without inwards. These rocks are formed at ordinary temperatures, and the associated water is of atmospheric origin. In the Endogenetic class are included the igneous rocks (along with certain pneumatolytic and hydrothermal types), and metamorphic rocks; and in the Exogenetic class come the rocks usually classed as sedimentary. The following gives a simplified statement of the classification :—

1. *Endogenetic Rocks.*

    (1) Igneous Rocks.
    (2) Igneous Exudation Products (due to pneumatolysis, metasomatism, etc.).
    (3) Thermodynamically-altered Rocks (Metamorphic Rocks).

2. *Exogenetic Rocks.*

    (1) Weathering Residues.
    (2) Detrital Sediments.
    (3) Solution Deposits.
    (4) Organic Accumulations.

[1] *Min. Mag.*, vol. xvii, 1914, pp. 55-86; also see A. Holmes, *Petrographic Methods and Calculations*, 1921, pp. 9-11.

PART I

THE IGNEOUS ROCKS

CHAPTER II

FORMS AND STRUCTURES OF IGNEOUS ROCKS

INTRODUCTION—The igneous rocks are those which have solidified from a molten condition. The most obvious and unequivocal examples are the lavas which have been poured out from present-day volcanoes. By comparison of their characters with those of certain igneous rocks intercalated in the geological column, it can be shown that the latter are the lavas of ancient volcanoes, the stumps of which can also sometimes be recognised.

Lavas are *extrusive* forms of molten magma which have been poured out on the surface of the earth. But for each surface eruption there must have been a vastly greater amount of subterranean activity. Hence there are *intrusive* or *injected* forms of igneous rock which are only exposed at the surface through subsequent denudation or earth movement.

There is, then, a broad distinction between extrusive and intrusive igneous rocks. The extrusive rocks are *erupted* to the surface ; the intrusive rocks are *irrupted* into the crust. In most cases it is easy to make the distinction in the field. The extrusive rocks, which have been exposed to the air, have lost most of their included gases, the escape of which has produced slaggy, cindery surfaces, vesicularity, and other characteristic structures, especially in the upper parts of the flows. Their rapid cooling is indicated by fine grain and the frequent presence of glass. Flow structures are also prevalent. On the other hand, intrusive rocks rarely show

13

vesicularity, or contain glassy matter, and they are usually much coarser in grain than the lavas. Furthermore, they alter the adjacent rocks on all sides, whereas in lavas only a slight amount of induration or baking is perceivable on the underside of the flows.

The present chapter deals with the forms and structures of igneous rocks, which depend largely on the mode of extrusion or intrusion.

FORMS OF IGNEOUS ROCKS

LAVA FLOWS—Lavas are emitted either from individual cones, like Vesuvius or Etna, or from fissures, such as are exemplified by some Icelandic eruptions. They usually form tabular bodies, of wide areal extent in proportion to their thickness, and elongated in the main direction of flow.

FIG. 2.—ERUPTION OF VISCOUS LAVA. A PUY.

The Grand Sarcoui, Auvergne. After Scrope. The central dome-shaped mass has heaped itself up around the vent, and is flanked by cinder cones.

The form of a lava flow depends chiefly on the fluidity of the magma, which, again, depends on its composition, and on the temperature of eruption. Thus basic lavas, such as basalt, are highly mobile, and flow for great distances; whereas acid magmas, such as rhyolite and trachyte, are sluggish in their flow and remain heaped up, often in steep-sided bulbous masses, about the orifice of eruption, as in some of the puys of Auvergne (Fig. 2). Basalt lavas have been described as ousting rivers from their beds, and producing lava-falls instead of waterfalls (Hawaii). One of the great basalt flows from the Laki (Iceland) fissure in the eruption of 1783, had a length of 56 miles, with an area of 220 square miles and a volume of 3 cubic miles (Fig. 3). Some special structures of lavas are dealt with later (p. 34).

PYROCLASTIC DEPOSITS—The intermittently explosive action that takes place in volcanic eruption produces a fragmental type of igneous material. The crust that forms over the lava column in a period of quiescence may be blown to pieces by a renewal of activity, and the fragments may be distributed in and about the crater as a mass of *agglomerate*. *Lapilli* are small cindery fragments between the size of a walnut and a pea, which are ejected to greater distances than the large fragments forming agglomerate. Finest of all are the dust-like particles which are produced by sudden explosion in liquid lava. These form *volcanic dusts* or *sands*, which may be deposited near the volcano or over the surrounding country, and may consolidate to form beds of *volcanic tuff*. All these ejectamenta may be mixed with foreign fragments torn from the sides of the volcanic vent, or with sand and mud co-deposited with the volcanic material. Thus beds of more or less pure volcanic debris

FIG. 3.—FISSURE ERUPTION.
Bird's eye view of basalt lava, 1783, Laki, Iceland. Fissure marked by line of small cones. See text, p. 14 for dimensions of flow. From Knebel-Reck, *Island*, 1912, pl. xvi.

are produced, the characters of which, however, especially in tuffs, partake more of those of sedimentary than of igneous rocks.

INTRUSIONS AND THEIR RELATIONS TO GEOLOGICAL STRUCTURES—Intrusions are those masses of molten rock which have been injected between the layers of the earth's crust. The forms that they take depend primarily on the geological structure, and subordinately on their relations to the structural features, such as bedding-planes, of the rocks which they penetrate. Two main types of geological structure may be distinguished in this connection : one in which the strata remain more or less horizontal over wide areas, but in which, while mainly unaffected by folding, they are frequently broken by nearly vertical fractures due to tension ; the other comprising mountain belts characterised by intense folding, contortion, and fracturing along gently-inclined planes (thrust-planes).   The forms of intrusions in these two strongly contrasted structural types are widely different.

The other factor by which the forms of intrusive bodies are governed is their attitude to the main structural planes of the rocks into which they are injected.   If the molten material has been guided by the bedding planes of the intruded rock, the resulting igneous body is said to be *concordant*.   On the other hand, the magma may break across the bedding planes, and then forms a *transgressive* or *discordant* mass.   The distinction between concordant and discordant is easiest to apply when the intrusions have penetrated strata which are more or less horizontal.   In that case concordant intrusions are approximately horizontal, and discordant masses approximately vertical.   When the strata were standing at high angles of dip before the intrusion of molten rock, the above criteria obviously become difficult to apply. This is because the terms concordant and discordant have reference only to the attitude of the intrusive body with respect to the structural planes of the invaded rock.   The true distinction, perhaps, is between intrusions which have had to lift the superincumbent cover of rock, and exert force in an approximately vertical · direction ; and those which have had to rend the rock apart, and thus exert force in an approximately horizontal direction, or which occupy fissures that are open, or ready to open, by reason of tension in the crust.   The distinction between concordant and discordant, however, is rarely difficult of application in the field.

It is possible for a concordant intrusion to depart for a

	Intrusions in Regions of Unfolded, Gently-folded, or Tilted Strata.	Intrusions in Regions of Highly-folded and Compressed Rocks.
Concordant.	Sill. Laccolith. Lopolith.	Phacolith. Concordant Batholith.
Discordant.	Dyke Cone-sheets. Volcanic Neck. Ring Dyke.	Discordant Batholith. (Stocks, Bosses). " Chonolith."

time from its typical attitude, and break transgressively across the structural planes, to resume concordance at a higher or lower level. Similarly a discordant intrusion may, for a part of its course, run along a main structural plane.

Intrusive bodies may thus be classified as shown in table on opposite page.

FORMS IN UNFOLDED REGIONS

Sills—Sills are relatively thin tabular sheets of magma which have penetrated along approximately horizontal bedding planes (Fig. 4). They show nearly parallel upper and lower margins for considerable distances; but as they thin out in distance the shape is flatly lenticular. The thickness may vary from a few inches to many hundreds of feet. Sills spread to a distance which is dependent on the hydro-

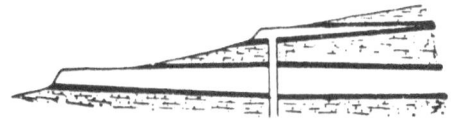

FIG. 4.—SILLS.

The sills (unshaded) have been injected into a series of sandstones (dotted) and shales (black). Both intrusions have been fed from the dyke (text, p. 18), and have penetrated along the shale beds.

static force with which they are injected, their temperature, degree of fluidity, and the weight of the block of strata which they have to lift in order to make room for themselves. Since basic magmas are more fluid than acid, the rocks composing the more extensive sills are usually dolerites and basalts. Some of the best examples of sills in this country are the quartz-dolerite and teschenite intrusions of North Britain. The largest of these is the Great Whin Sill (quartz-dolerite), which traverses five counties in the North of England, and covers an area of at least 1500 square miles.

The source of the quartz-dolerite sills is the series of great east-west dykes traversing the above region, which have risen along lines of weakness, fissures, and faults. While a connection between dyke and sill can occasionally be demonstrated, the dykes are often seen to cut through a sill and

2

reach higher horizons in the strata, where another sill may be thrown off. Nevertheless, the sill may have proceeded from the dyke that traverses it, for the latter may give off a sill at one horizon, the spread of which may be checked by congelation, or because the energy of the intrusive impulse may be insufficient to cause it to spread further. Then, if the hydrostatic force is still maintained, the dyke may break through the sill at its weakest point, namely, near the orifice

Drakensberg
Volcanics
Cave Sandstone.
Red Beds
Molteno Beds

Beaufort Beds.

Ecca Beds.

Dwyka Conglomerate
Table Mountain
Sandstone.

Granite, etc.

FIG. 5.—SILLS IN THE KARROO, SOUTH AFRICA.

Diagrammatic section showing the distribution of sills in the Karroo formations. After A. L. du Toit. Note transgressive connecting portions and sills passing into dykes (text, p. 19).

of injection where the magma may still be liquid, and resume its course along the original fault or fissure.

Sills are prominent features in many plateau countries. One of the best examples is the Karroo region of South Africa where, according to Dr. A. L. du Toit, dolerite intrusions, mostly of sill character, penetrate the strata over an area of 220,000 square miles.[1] In South Africa, as in Scot-

[1] *Trans. Geol. Soc. S. Africa*, xxiii, 1920, p. 2.

land, the sills tend to spread along the weaker horizons in the strata, along partings of shale within sandstone, or along coal seams. They are disinclined to enter, or alter their forms in, the harder rocks in which bedding planes may be few or absent. Sills frequently break across from one level to another along cross joints or other fissures which temporarily offer an easier path, but these discordances are always quite subordinate in extent to the concordances (Fig. 5).

LACCOLITHS—A magma of considerable viscosity injected into stratified rocks does not spread very far, but tends to heap itself up about the orifice of irruption. Thus a bun-shaped mass of igneous rock is formed which has a flat base

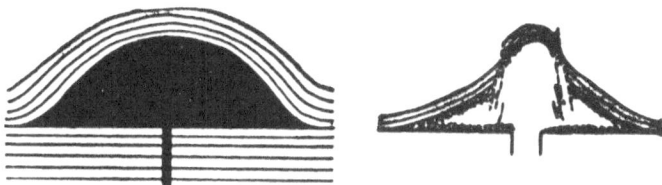

FIG. 6.—LACCOLITHS.

A. Typical laccolith.
B. Upfaulted laccolith or bysmalith. After Paige. Dotted portion represents stiffened magma at contacts (text, p. 20).

and a domed top ; the strata above it are lifted up in the form of an inverted bowl. This is the type of intrusion which was first described by G. K. Gilbert as a laccolith (Fig. 6 A).[1] Put in another way, the laccolithic form is produced if the supply of magma from beneath is greater than can be accommodated by lateral spreading. Laccoliths are occasionally circular, or more often elliptical, in ground plan, according as the supply is from a cylindrical vent or from an elongated fissure. Small sills may proceed from their margins, and dykes may occupy tension cracks in the dome of stretched strata above.

The effects of progressive increase of viscosity during the injection of a laccolith, and its stiffening due to cooling at

[1] *Geol. of the Henry Mts.*, 1877, p. 19.

the contacts, may cause a steepening of the lateral margins of the mass almost to verticality; and if the irruptive impulse continues unchecked, this will pass into fracture.[1] The overlying cylinder of strata will be, as it were, punched upward, and the igneous mass will be surrounded by arcuate faults. This is the type of intrusion which Iddings has called a *bysmalith* (Fig. 6 B).[2] A bysmalith is therefore partly a cross-cutting or discordant body.

Keyes holds that preliminary faulting and arching of the strata during mountain-building processes are necessary pre-requisites to the production of laccoliths.[3]

LOPOLITHS—The name *lopolith* (Gr. *lopas*, a basin or flat earthen dish) has been given to massive intrusions of basic

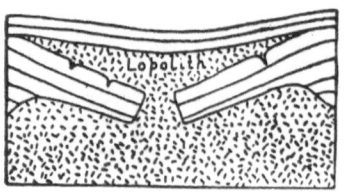 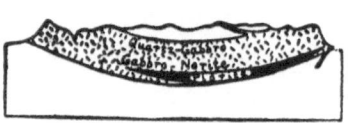

| A | FIG. 7.—LOPOLITHS. | B |

A. Theoretical section through a lopolith.   After Grout (text, p. 20).
B. Section of part of the Insizwa lopolith.   After A. L. du Toit (text, p. 21).   Length of section about 14 miles.   This is a differentiated intrusion.

rocks which are generally concordant, have a lenticular shape, and are centrally sunken like a saucer or basin.[4] The thickness is approximately one-tenth to one-twentieth of the diameter (Fig. 7 A). The type example is the Duluth (Minnesota) gabbro mass, which has a diameter of 150 miles, and a maximum thickness estimated at 50,000 feet. Its outcrops enclose an area of 15,000 square miles, and its volume is believed to be of the order of 50,000 cubic miles.

Lopoliths thus differ from laccoliths in their enormously greater size, in their form, and in the pronounced sagging of

[1] S. Paige, *Journ. Geol.*, 21, 1913, p. 541.
[2] *Journ. Geol.*, 6, 1898, p. 704.
[3] *Bull. Geol. Soc. Amer.*, 29, 1918, p. 75.
[4] F. F. Grout, *Amer. Journ. Sci.*, 46, 1918, p. 516.

their bases. It has been suggested that the famous " nickel-eruptive " of Sudbury, Ontario, which has a basin shape, is also a lopolith. The Insizwa gabbro of South Africa is part of a great basin-shaped mass which originally covered 700 square miles and had a volume of 300 cubic miles[1] (Fig. 7 B). The great Bushveld igneous mass of the Transvaal has also been cited by Grout as a lopolith, but Daly and Molengraaff have shown that it is not a simple mass, but a group of igneous units, which, pending fuller knowledge, should be termed a " complex."[2]

DYKES—The intrusion of magma into more or less vertical fissures which cut across the bedding or other structures of the invaded rock, results in the formation of *dykes*. A dyke (Scots, *dike*, a wall of turf or stone), as the name

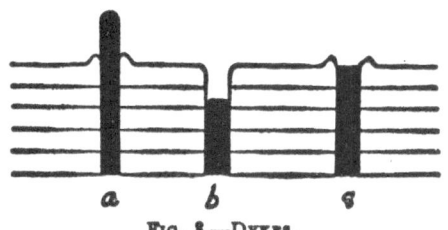

FIG. 8.—DYKES.
Illustrating their modes of weathering (text, p. 22).

indicates, is thus a narrow, elongated, parallel-sided wall of igneous rock. The thickness may vary from an inch or two to many hundreds of feet, but the great majority of dykes are probably less than 10 feet thick. Similarly, the length of a dyke may vary from a few yards to many miles. Some of the great east to west dykes of quartz-dolerite which cross the Midland Valley of Scotland have lengths of the order of 30 to 40 miles, and are upwards of 100 feet thick. On the other hand, the thousands of dykes in Arran and other centres of Kainozoic volcanic activity in the west of Scotland, are usually only a few feet thick, and can seldom be traced for more than a few hundreds of yards.

Dykes are frequently more resistant to erosion than the

[1] A. L. du Toit, *Trans. Geol. Soc. S. Africa*, 23, 1920, p. 19.
[2] *Journ. Geol.*, 32, 1924, p. 2; R. A. Daly, *Bull. Geol. Soc. Amer.*, 39, 1928, pp. 703-68.

enclosing rocks, and tend to project as walls at the surface
(Fig. 8 A); sometimes, however, they weather more easily,
and are then excavated out to form a trench (Fig. 8 B).
Occasionally they bake and harden the adjacent walls so
that the indurated ridges stand out on either side of the
dyke (Fig. 8 c).

Dykes tend to occur as systems or swarms which are parallel to one direction, or are radial to a centre. Thus the
dykes associated with Kainozoic volcanic activity in the west
of Scotland have a characteristic north-west to south-east

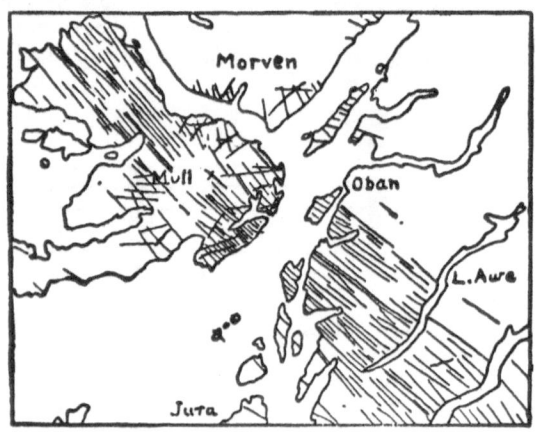

FIG. 9.—DYKE-SWARM, MULL.

Each line represents a group of from ten to fifteen dykes. The Mull
   swarm is of Kainozoic age. (From the *Mull Memoir*, 1924,
   p. 357.)

trend, and are disposed in distinct groups or *swarms*. The Mull
swarm of parallel dykes, crossing Mull and the Argyllshire
mainland from north-west to south-east, is about 10 miles
wide (Fig. 9).[1] As an example of a radial swarm may be
cited the Cheviot district, in which the dykes are disposed
radially about the central granite of Old Red Sandstone age.[2]

[1] *Mem. Geol. Surv. Scotland : Tertiary and Post-Tertiary Geology
of Mull, L. Aline, and Oban*, 1924, p. 359. Hereinafter cited as the
Mull Memoir.
[2] A. Harker, *Nat. Hist. Ign. Rocks*, 1909, Fig. 2, p. 26.

Dykes are symptomatic of regional tension in the crust within an area of igneous activity. They have to open up their way along fractures, for it is extremely rare to find open fissures of the width necessary for simple infilling. Owing to the existing tension the intrusion of magma, so to speak, touches off or relieves the tensile stress, the fissures opening up with a minimum expenditure of the energy of the intrusion. Thus the injection of dykes may take place with

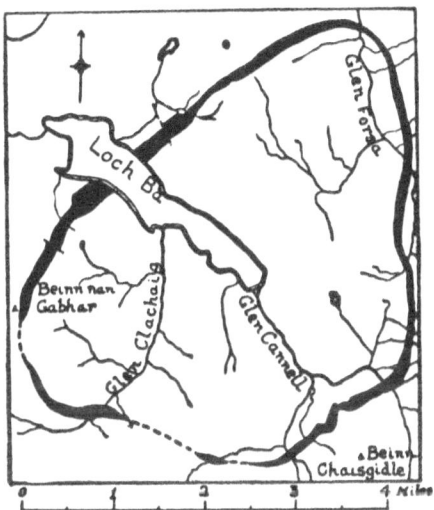

FIG. 10.—RING-DYKE OF LOCH BÀ, MULL.
See text, p. 23. The dyke consists of felsite. (From the *Mull Memoir*, 1924, p. 338.)

great rapidity, which is also to be inferred from their great linear extensions and narrow widths. Thus the north-west to south-east extension of the Kainozoic swarms of the west of Scotland indicates a regional tension operating from north-east to south-west in that area. Radial dyke systems may be ascribed to tensions developed around local centres of eruption and intrusion.

RING-DYKES AND CONE-SHEETS—A *ring-dyke* is a dyke of arcuate outcrop which, with full development, would have a

closed, ring-shaped outcrop.[1]   Ring-dykes may be arranged
in concentric series, with *screens* of country-rock separating
the individual members of the complex.   Some of the most
perfect examples of ring-structures are to be found in Mull
and Ardnamurchan among the Kainozoic intrusions (Fig. 10);
other good examples are those of the Old Red Sandstone
igneous episode of Lorne; and they have also been noted in
Iceland and in the Oslo (Kristiania) igneous province of
Norway.   Ring-dykes are usually thick, and may consist of
coarse-grained plutonic rocks.   Some of the more massive
ring-intrusions seem to have risen up arcuate fractures defining the margins of calderas or volcanic subsidences (e.g.

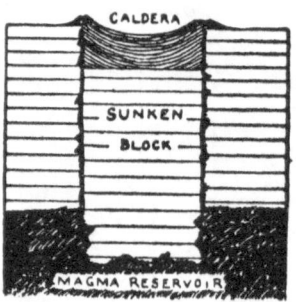

FIG. 11.—ORIGIN OF RING-DYKES.

Formation of a ring-dyke around a sunken block.   Volcanic eruption
and caldera formation at the surface (text, p. 24).   (From " The
Geology of Ben Nevis and Glencoe," *Mem Geol. Surv. Scotland*,
1916.)

those of Old Red Sandstone age in Lorne).   Fig. 11 shows
diagrammatically the mode of origin of ring-dykes.[2]

CONE-SHEETS (inclined-sheets, Harker), or cone-sheet complexes, are assemblages of inclined dyke-like masses with
arcuate outcrops, the members of which dip at angles of
30° to 40° towards common centres.[3]   Their arrangement suggests the partial infilling of a number of coaxial cone-shaped
fractures, with inverted apices united underground.   Cone-

[1] *Mull Memoir*, Chaps. XXIX, XXXII.
[2] "Geology of Ben Nevis and Glencoe," *Mem. Geol. Surv. Scotland*,
1916, p. 126.
[3] *Mull Memoir*, Chaps. XIX, XXI, and XXVIII.

sheets have hitherto only been recognised among the Kainozoic igneous rocks of Skye, Mull, and Ardnamurchan, where
they are extremely abundant. The phenomenon was first
recognised in Skye by Dr. A. Harker, who applied the name
inclined sheets.[1] Individual cone-sheets may reach 30 feet

FIG. 12.—CONE-SHEETS, SKYE.

Section illustrating the distribution of cone-sheets in the gabbro region
of the Cuillin Mountains. The inclination of the cone-sheets increases towards the central area. Broken line indicates the base of
the gabbro. Length of section, 12 miles. After Harker (text,
p. 25).

to 40 feet in thickness; but, in a Mull example, the aggregate thickness of a complex of these intrusions exceeds
3000 feet. Fig. 12, adapted from Harker (*op. cit.*, p. 368),
indicates the arrangement of cone-sheets in the gabbro
region of the Cuillin Mountains of Skye.

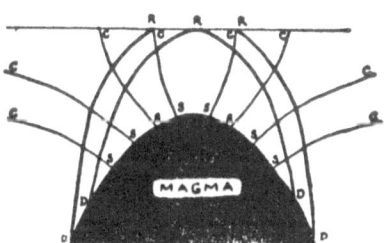

FIG. 13.—ORIGIN OF CONE-SHEETS AND RING-DYKES.
After E. M. Anderson (text, p. 25).

The origin of cone-sheets and ring-dykes has been explained
by E. M. Anderson as a dynamical consequence of magmatic
intrusion and withdrawal.[2] In Fig. 13 the lines marked CS
indicate the traces of cone-shaped fractures which would be

[1] *Tert. Ign. Rocks of Skye*, 1904, Chap. XXI, p. 364.
[2] *Mull Memoir*, p. 11.

developed by increase of pressure in the magma reservoir. It will be noted that the fracture planes steepen in depth, and the cone-sheets produced by their infilling should be more steeply inclined in the central parts of their area of intrusion, which is actually the case in Skye (see Fig. 12) and Mull. If, on the other hand, the magmatic pressure lessens, surfaces of fracture would develop the traces of which are indicated by the lines RD. These surfaces correspond more to planes of maximum shearing stress rather than to tension cracks. Their infilling with magma would give rise to the phenomena of ring-dykes. The rocks enclosed within the fractures would tend to sink downward, and this would

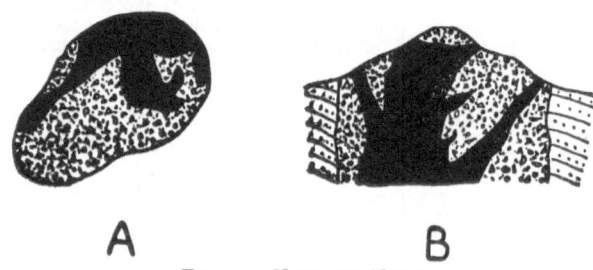

A   B

FIG. 14.—VOLCANIC NECKS.

Diagrammatic sections through a volcanic neck. Plan (A) and section (B) through Dumgoyn (text, p. 26). After Sir A. Geikie, *Ancient Volcanoes of Great Britain*, 1897, vol. i, pp. 395, 400. A basaltic plug (black) intrusive into agglomerate (stippled).

explain, partly at least, the greater width of ring-dykes as compared to cone-sheets.

VOLCANIC NECKS—These are igneous masses which seal up the vents of ancient volcanoes. They may completely occupy the cylindrical channel or may be intrusive around and into the agglomerate which partly fills it. The plan and section of Dumgoyn (Fig. 14 B), the famous volcanic vent at the western end of the Campsie Fells in the Midland Valley of Scotland, illustrate the latter case. Volcanic necks may throw off sills and dykes into the surrounding strata, and may, in exceptional cases, pass laterally into lava flows.

FORMS IN FOLDED REGIONS

PHACOLITHS—The forms of igneous intrusions in folded regions are not so simple and regular as those in unfolded regions, because the structure of the invaded rocks is much more complicated, and the fracture planes much less regular. One of the simplest types is that described by Harker as a *phacolith*.[1] In folding the crests and troughs of the folds become regions of weakness and tension, whereas the middle limbs are compressed ; so that igneous material, when present, will tend to find its way into the crests and troughs, and will there exhibit doubly-convex, lens-like forms (Fig. 15). Most phacoliths are of comparatively small dimensions, and corre-spond in this respect to laccoliths among intrusions in un-folded regions. They are, however, fundamentally different from laccoliths, as their location and shape are determined

FIG. 15.—PHACOLITHS.
Diagram illustrating location of phacoliths in anticlines and synclines
After Harker (text, p. 27).

by the conditions of folding, and their intrusion is not, as in the case of laccoliths, the cause of the attendant folding. Transitions will probably be found between phacoliths and the concordant type of batholith (see p. 28).

CHONOLITHS.—The term *chonolith* (= mould-stone ; i.e. a magma filling its chamber as a metal fills the mould) has been used by Daly to designate those irregularly-shaped intrusive bodies which do not fall under any of the yet recognised categories of form.[2] A chonolith is defined as an igneous body injected into dislocated rock of any kind, of irregular shape and relations, and composed of magma passively squeezed into an underground orogenic chamber, or actively forcing the invaded rocks apart. Chonolith is thus designedly

[1] *Nat. Hist. Ign. Rocks*, 1909, p. 77. Fine examples of phacoliths have been described by Th. Vogt from northern Norway, "Sulitelma-feltet Geologi og Petrografi," *Norske Geol. Undersök*, No. 121, 1927. English summary, pp. 449-560.
[2] *Ign. Rocks and Their Origin*, 1914, p. 84.

an " omnibus " term, but it will serve a useful purpose in accommodating unclassified bodies of igneous rocks until such time as further categories of form are recognised.

BATHOLITHS.—Batholiths (Fig. 16) are the greatest bodies of igneous rock known. In detail they usually have trans- gressive relations to the rocks they invade, but as they are frequently elongated parallel to the strike, they are in one sense of concordant habit. Parts of the roof of a batholith may occasionally be recognised (roof-pendants), but as the mass often apparently broadens down to unknown depths, the nature of the base is purely a matter of speculation. So

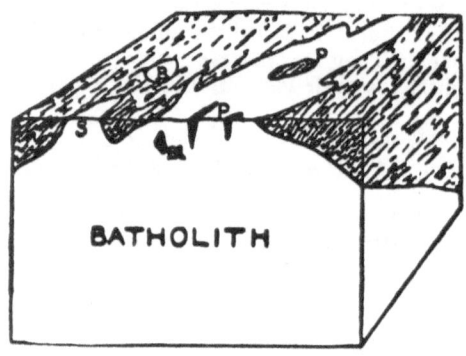

FIG. 16.—RELATIONS OF BATHOLITHS TO COUNTRY ROCKS.
See text, p. 28.   S, stock; B, boss; P, roof-pendants; Bl, sunken block. After Daly.

impressed was Professor R. A. Daly [1] by the lack of evidence of a true floor to batholiths that he separated them off from all other intrusive masses as *subjacent* to, or underlying, the crust, as opposed to other bodies of magma which are in- jected between the strata, and thus have a true base of country-rock. According to Daly batholiths are character- ised by the following features : [2] They are located in moun- tain-making zones, and are elongated parallel to the tectonic axes of the folded belts. Their intrusion follows more or less closely an antecedent period of mountain building. Towards

[1] *Ign. Rocks and Their Origin*, 1914, p. 62.   [2] *Ibid.*, p. 90.

the adjacent rocks they have transgressive relations ; the roof is irregularly dome-shaped, the lateral walls steeply inclined and relatively smooth. The igneous mass appears to enlarge downward, no floor is visible, and it has the appearance of having replaced the invaded formations. The composition of batholiths is usually granitic or granodioritic.

Many batholiths are of enormous dimensions ; amongst the largest are the Alaska-British Columbia granite mass with an extension of about 1250 miles and a width in places of 50 miles ; and the Patagonian batholith of South America, which has a length of 700 miles, with a greatest width of 70 miles. Professor Daly fixes the lower limit of size at 100 square kilometres (40 square miles). *Stocks* are irregular masses of batholithic habit of smaller dimensions than the above, and the term *boss* is applied to those stocks in which the outcrop is approximately circular. Stocks and bosses are usually offshoots from a larger batholithic mass (Fig. 16).

The mode of origin and emplacement of batholiths are questions which still afford matter for controversy. E. Suess, the originator of the term, at first thought batholiths represented the infilling of open spaces in the crust which were caused by folding and thrusting.[1] There is no doubt that batholiths tend to follow the major planes of fracture, weakness, and discontinuity in the earth's crust, as has been emphasised by Iddings,[2] and especially by Cloos.[3]

Controversy regarding the mode of emplacement of batholiths now centres mainly around the question whether the igneous mass made room for itself by the destruction and replacement, or by the displacement, of the invaded rocks. The French school of petrologists, as also the late Professor G. A. J. Cole, believed that magma was capable of melting its way through the crust, and incorporating the melted material into itself. This process, however, would require an enormous amount of localised super-heating of the magma, of which there is no evidence ; and further, as the composition of batholiths is remarkably uniform on the whole, the

---

[1] *The Face of the Earth*, English trans., vol. i, pp. 168-9.
[2] *The Problem of Volcanism*, 1914, pp. 203 *et seq.*
[3] *Das Batholithenproblem, Fortschr. d. Geol. u. Pal.*, 1, 1923, p 54 *et seq.*

FIG. 17.—SECTION THROUGH BAVARIAN BATHOLITH. (See text, p. 30.) After Cloos.)

idea of wholesale assimilation of foreign material is improbable, at any rate, in the upper levels of the crust.

Daly [1] and Barrell [2] elaborated the idea of mechanical dislodgment or stoping of blocks from the roof of the intrusion by unequal heating, veining, and magmatic wedging. The resulting fragments are supposed to sink through the magma, and to be dissolved in depth, although at a late magmatic stage some may be retained near the contacts. This process is supported by the irregular and embayed margins of many batholiths, and by the frequent presence of blocks of all dimensions enclosed within the intrusion.

Iddings and Cloos (*op. cit., supra*), however, regard stoping as merely a subordinate and incidental factor in the emplacement of batholiths. The necessary space for their development has been gained by the displacement of the adjacent rocks which have been thrust aside, lifted, and even partly forced downward, concurrently with the folding and thrusting action of mountain-making forces. Both Iddings and Cloos question the hypothesis that batholiths are bottomless. In the typical granite batholiths of South Bavaria Cloos has shown that the masses lie between, and not under, the older rocks which they have invaded. The same rocks which abruptly end off against the granite contacts are found again beneath the mass, and continue unchanged (Fig. 17).[3]

[1] *Amer. Journ. Sci.*, 15, 1903, pp. 269-98 ; *Ign. Rocks and Their Origin*, 1914, Chap X.
[2] *Prof. Paper*, No. 57, *U.S. Geol. Surv.*, 1907, pp. 151-74.
[3] Cloos, *op. cit.*, p. 14 *et seq.*

MULTIPLE INTRUSIONS—When a magma has been injected along a single channel in two or more successive pulses, the result is a *multiple* intrusion (Fig. 18). Many multiple sills and dykes have been described, especially from the Kainozoic igneous episode of the west of Scotland. The criteria for distinguishing multiple intrusions are: (*a*) one or more interior contacts, where later injections have been cooled against previous ones; and (*b*), uniformity of composition throughout the complex.[1] Succeeding injections may occur along one or both margins of the earlier intrusion, or may pass along planes of weakness in the interior (see under Composite Intrusions).

COMPOSITE INTRUSIONS—When magmas of different composition avail themselves of the same channel of injection they give rise to the phenomenon of *composite* intrusion. If

FIG. 18.—MULTIPLE SILL.

Triple multiple sill, consisting of two members, 1 and 2. Chilled edges shown by dots.

there has been a sufficiently long interval between the successive intrusions, the later one will be chilled against the earlier. Often, however, it is clear that the later injection has taken place before the earlier one had time to cool. In this case there is a good deal of commingling of the magmas, with enclosure of fragments and hybridisation, along the interior contacts (see p. 32). As in multiple intrusions, the successive injections may pass along one or both contacts of the first intrusion, or along interior planes of weakness. Hence composite and multiple intrusions may be double,

[1] This criterion is rejected by H. H. Thomas and E. B. Bailey (*Mull Memoir*, 1924, p. 32). According to them, multiple intrusions must show interior contacts, but may be composed of rocks of different composition. Composite intrusions, on the other hand, are composed of members of different composition which do not show chilled contacts against one another.

triple, or even quintuple. Dr. W. R. Smellie [1] has shown that thin dykes develop a plane of weakness near their centres (Fig. 19 A), whereas thicker dykes develop two planes of weakness, each parallel to, and within one or two feet of,

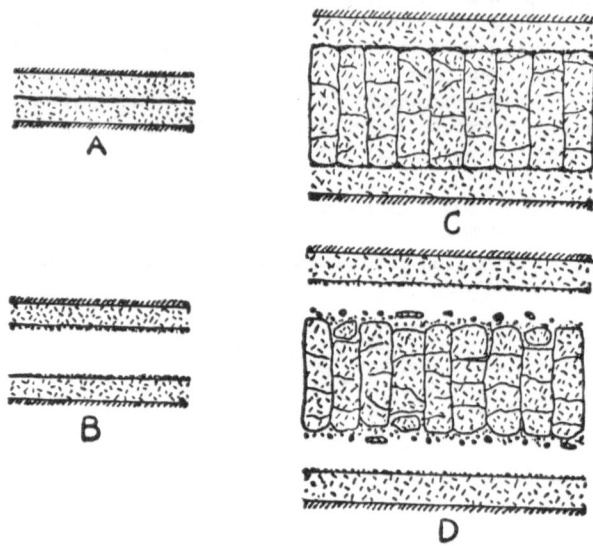

FIG. 19.—COMPOSITE DYKES AND SILLS.

A. Thin intrusion (dyke or sill) with median plane of weakness.
B. The same, showing subsequent intrusion along median plane, with enclosure of xenoliths along both margins, and formation of triple composite intrusion.
C. Thick intrusion with two planes of weakness developed near the contacts Interior with rough columnar jointing.
D The same, showing subsequent intrusion along planes of weakness, with enclosure of xenoliths along margins, and formation of quintuple composite sill.

the contact planes (Fig. 19 c). Hence, if another magma seeks to pass along the same channel it will probably choose these planes of weakness in preference to the tightly-welded contacts, giving in the one case a triple composite dyke

[1] *Trans. Geol. Soc. Glasgow*, xv, part 2, 1914, p. 125.

(Fig. 19 B), and in the other, a quintuple composite dyke (Fig. 19 D), each with two members of differing composition. Sills may be split up in exactly the same way. It is obvious that belts of xenolithic material may be formed along the interior contacts by this mode of injection.

The so-called *cedar-tree laccoliths* are multiple or composite sills of many successive injections, in which the various members fray out individually in the bedding planes of the adjacent strata.

DIFFERENTIATED INTRUSIONS—In a sufficiently large intrusion of uniform magma, the process of differentiation (see p. 148) may give rise to parts of contrasted composition before the final consolidation takes place. Thus *differentiated* intrusions are formed, in which the parts of differing composition usually pass gradually into one another. As size and slow cooling are factors in the promotion of differentiation, it is clear that the larger intrusive masses, i.e. the greater sills, laccoliths, lopoliths, and batholiths, will most often be differentiated (see Fig. 48, the Lugar sill, for an example of an intrusion which is at once composite and differentiated).

STRUCTURES OF IGNEOUS ROCKS

STRUCTURE AND TEXTURE—Under the term *structure* are included certain *large-scale* features, such as the blocky or ropy surfaces of lavas, pillowy structures, flow-banding, jointing structures, and the potential fractures of rift and grain. Structure also denotes some *small-scale* features which are due to the juxtaposition of more than one kind of textural aggregate within a rock, such as amygdaloidal and spherulitic structures. *Texture*, on the other hand, indicates the intimate mutual relations of the mineral constituents and glassy matter in a rock made up of a uniform aggregate. Textures and small-scale structures are treated together in a later chapter; the present section will deal only with large-scale features, including thereunder vesicular and amygdaloidal structures,[1] block lava and ropy lava, pillow structure,

---

[1] Strictly speaking, these are small-scale structures. They are dealt with here because they are important in the explanation of block and ropy lava, pillow lava, etc.

and flow-banding, which are due mainly to magmatic forces acting from within, and other structures which are mainly due to jointing and parting (p. 39).

VESICULAR AND AMYGDALOIDAL STRUCTURE—Most lavas are heavily charged with gas, which escapes as soon as the pressure is diminished by their eruption at the surface. The escape of the gases distends the molten material with the production of cavities, bubbles, or vesicles, which may be spherical, elliptical, cylindrical, or irregular, in shape. The name *slag* or *scoria* is applied to lava in which the gas cavities are very numerous and irregular in shape. *Pumice* is produced in an extreme stage of distension by escaping gases, when the lava forms a kind of molten froth. Long cylindrical or tubular vesicles (pipes) are occasionally found at the base of a flow rising perpendicularly from the lower margin. These may be due to gases disengaged from underlying sediments by the heat of the lava. In certain doubtful cases pipe vesicles (and amygdales) have been found stratigraphically useful in determining whether a particular bed of lava was or was not " right way up." [1]

Amygdales are the infillings of vesicles by secondary minerals, and are so-called because their shapes sometimes suggest a fanciful resemblance to almonds. Lavas containing amygdales are said to have *amygdaloidal* structure, and the rocks themselves are called *amygdaloids*. The infilling minerals may be calcite, various forms of silica, zeolites, or indefinite hydrated ferro-magnesian silicates which go collectively under the name of " green earths." Most amygdales are to be regarded as the products of the final exudations of the lavas themselves, or of the magma which gave rise to them.

BLOCK LAVA AND ROPY LAVA—Two very different appearances may be presented by lava flows. Sometimes the surface is covered with a mass of rough, jagged, angular blocks of all dimensions, like a sea of clinkers, which, during the flow of the lava, are borne along as a tumbling, jostling mass (Fig. 20). There are many names for this phenomenon ; *block lava* is, perhaps, the simplest. The name *aa*,

---

[1] "Geol. of Knapdale, Jura, and North Kintyre," *Mem. Geol. Surv. Scotland*, 1911, p. 69.

applied by Hawaiian natives to lava fields of this type, has been frequently used by vulcanologists; in Iceland they are termed *apalhraun*.[1] Jaggar has given them the systematic name *aphrolith* (= foam-stone).[2]

On the other hand, very mobile lavas solidify with much

FIG. 20.—BLOCK LAVA, VESUVIUS.
The flow of 7 Feb., 1906. Drawn from photograph in F. A. Perrett's *The Vesuvius Eruption of 1906*, 1924.

smoother surfaces, often highly glazed, which, in detail, exhibit wrinkled, ropy, or corded forms, similar to those displayed by flowing pitch (Fig. 21). The surface is also diversified by low domes or lava blisters a few metres in

[1] Icel. *apal* = grey moss. An allusion to the grey moss which quickly covers this type of lava. *Hraun* = lava.
[2] *Journ Wash. Acad. Sci.*, 7 1917, p. 277

diameter, which show characteristic radial cracks. This is the *ropy* type of lava ; called *pahoehoe* in Hawaii, *helluhraun* [1] in Iceland, and *dermolith* (= skin-stone) by Jaggar (*op. cit.*). The block and ropy types of lava are best developed in basalts, and may occur side by side in the same flow.

The vesiculation of the two types of lava differs considerably. In block lava the cavities are large and irregular in shape ; in ropy lava they are smaller, much more numerous, and of regular form, being mostly spherical. The total volume of vesicle space per unit volume is greater in ropy

FIG. 21.—ROPY LAVA, VESUVIUS.
Lava on crater floor, 26 Aug., 1918.   Drawn from photograph in
F. A. Perrett's *The Vesuvius Eruption of 1906*, 1924.

than in block lava. The only chemical difference between blocky and ropy basaltic lava is that the proportion of ferrous iron oxide (FeO) to ferric oxide (Fe_2O_3) is uniformly greater in the ropy form.[2] The latter, too, is always less crystalline than the blocky lava, and usually occurs in much smaller flows.

According to Washington (*op. cit.*) these facts are explained by the theory that ropy lava issues at a higher temperature than block lava, but with a much smaller content of gas.

---

[1] Icel. *hellu*, smooth.
[2] H. S. Washington, " The Formation of Aa and Pahoehoe," *Amer. Journ. Sci.*, 6, 1923, pp. 409-23.

The gas soon escapes and the flow quickly congeals with a minimum amount of crystallisation. Block lava, on the contrary, issues at a lower temperature, but is so highly gas-charged that it is initially even more fluid than the ropy type. Under these conditions crystallisation begins early and proceeds with rapidity. The escape of gas becomes increasingly rapid and even violent towards the moment of consolidation, but the remaining liquid is always gas-saturated, and thus in a condition favourable to crystallisation.

It follows from this explanation that block and ropy lava must be regarded as the end-members of a series of lava forms which are connected by transitional varieties. Perrett, indeed, hints at the possibility of instituting a regular subdivision of the range of form between the two extremes somewhat on the lines of the Rossi-Forel scale for earthquakes.[1]

PILLOW STRUCTURE (*Ellipsoidal Structure*).[2]—This is a peculiar structure occurring mostly in basic lavas, and especially in the soda-rich basaltic types known as *spilites*, in which the lava exhibits the appearance of a pile of small masses which have been variously compared to pillows, bolsters, sacks, and cushions. The pillow generally has a vesicular crust and occasionally a glassy skin. It frequently exhibits flow-banding, often by lines of vesicles concentric with its surface. The pillows often make indentations upon their neighbours as if they had been soft at the time of their formation. The inter-spaces between the pillows are filled sometimes with a breccia cemented by secondary minerals, but more often with radiolarian chert and impure siliceous limestone. The masses are frequently elongated, like bolsters, and there may be a pronounced parallelism between their longest axes. Lastly, the pillows may be connected one with the other by short tubes or necks, or along their sides.

All transitions between true pillow lavas and the wrinkled,

[1] " The Vesuvius Eruption of 1906: Study of a Volcanic Cycle," *Carn. Inst. Wash. Publ.*, No. 339, 1924, p. 75.
[2] The term " pillow " was first used in connection with this structure in the description of the variolitic rocks of Mt. Genèvre by G. A. J. Cole and J. W. Gregory, *Q.J G.S.*, xlvi, 1890, p. 312.

ropy, bulbous forms of pahoehoe have been observed. A spherical or ellipsoidal jointing occasionally produces masses which simulate pillows, but these do not possess the distinctive features referred to above, and are usually easily distinguished.

The association of pillow lavas with marine sediments has led to the view that the structure is due to the contact of molten lava with sea water. But extrusion into rain-soaked air, or beneath ice-sheets, or intrusion into soft, water-logged sediments, may induce sufficiently rapid cooling to produce the structure.

Pillow structure is only found in freely-flowing basaltic lavas which retain a high degree of liquidity through a lengthy period of cooling, but nevertheless develop considerable viscosity as they approach the solidifying point.[1] In the declining stages of a lava flow with these characters a crust forms on the surface, and the flow can only be continued by the escape of liquid lava through fissures in this crust. Thus, for one large flow there may be substituted a multitude of small flows, each of which will form an elongated bulbous mass. A tough elastic skin forms on the surfaces of these small flows, and the internal pressure of the lava will expand this skin until rupture occurs, with the formation of a still smaller extrusion which quickly congeals into a pillow. The expansion of each pillow produces a pseudo-flow structure parallel to the outer surface, concentric arrangement of vesicles, and tangential arrangement of microlites and phenocrysts. The glassy skin and vesicular crust are due to rapid cooling induced by contact with water or other moisture-laden medium.

The interspaces are filled with fragments spalled off from the pillows, or with sediments of the sea floor. The constant association of pillow lava with radiolarian chert has been explained by Dewey and Flett as due to the discharge of solutions rich in silica from the lava into the sea, thereby providing conditions favourable to the rapid multiplication of siliceous organisms such as radiolaria.[2]

The actual formation of pillow lava was observed by Dr.

---

[1] J. V. Lewis, *Bull. Geol. Soc. Amer.*, 25, 1914, p. 646.
[2] "British Pillow Lavas," *Geol. Mag.*, 1911, pp. 244-5

Tempest Anderson, where the lava from the volcano of Matavanu in Savaii (Samoa) ran into the sea.[1]   On flowing into the water ovoid masses of ropy lava were seen to swell and crack into a sort of bulb with a narrow neck, and this would increase until it became a lobe as large as a sack or pillow. The connecting necks were sometimes long, but most often so short that the freshly-formed lobes were heaped together.

Ancient pillow lavas have been found in many localities and in formations belonging to many geological periods. Some of the finest examples in the British Isles are to be found in the Arenig lavas of the Ballantrae district of Ayrshire.[2]

FLOW STRUCTURE—No lava is ever quite homogeneous during and immediately after extrusion. Layers and patches in it differ slightly in composition, gas-content, viscosity, and degree of crystallisation. In the process of flow these patches are drawn out into parallel lenticles, streaks, bands, and lines, which may be characterised by the development in various proportions of vesicles, spherulites, glass, microlites, crystals, and stony material of slightly varying composition manifested by slight differences in colour and texture. Acid and sub-acid lavas, such as rhyolites and trachytes, the magmas of which are extremely viscous, show flow structures in greatest perfection.

Banded structures are also to be found in plutonic rocks, due to the alternation of layers differing in mineral composition, texture, or both. This structure, however, may be due to a number of causes, of which magmatic flow is only one.[3] It is conceivable that the injection of an originally heterogeneous magma might result in the streaking out of the different parts into layers of different composition and texture.

JOINTING, SHEET, AND PLATY STRUCTURES—We now deal with structures characterised by fracture or parting, which

[1] *Q.J.G.S.*, lxvi, 1910, pp. 631-3.
[2] "The Silurian Rocks of Great Britain." *Mem. Geol. Sur.*, vol. i, Scotland, 1899, p. 431 (Pls. I, II, V, VI).
[3] See A. Holmes, *Petrographic Methods and Calculations*, 1921, pp. 365-6.

are impressed upon the rock mainly by forces acting from without.

JOINTS are divisional planes which are found in all kinds of igneous rock. In granites it is common to find three sets of joints, one more or less horizontal, the other two vertical and perpendicular to each other. If these three systems of partings are more or less equally spaced, the fracture planes give rise to a structure of cuboidal blocks resembling a gigantic wall (mural jointing, see Fig. 22). The horizontal

FIG. 22.—MURAL JOINTING IN GRANITE, ARRAN.

Near summit of Goatfell. Drawn from photograph in " The Geology of North Arran, South Bute, and the Cumbraes," *Mem. Geol. Surv.*, *Scotland*, 1903.

joint-planes are sometimes so closely spaced as to produce a *sheet* structure. The sheets are commonly thinner as the surface of the ground is approached, and they usually show some degree of parallelism to the surface.[1]

In many igneous rocks, however, the jointing is much less regular. The divisional planes may be curved or undulating. This feature is commonly seen in the Lower Carboniferous

[1] For a full discussion of sheet structure see T. N. Dale, *The Commercial Granites of New England : U.S. Geol. Surv. Bull. No.* 738, 1923, pp. 26-36.

basalt lavas of Scotland.[1] It often exists as potential fracture planes which are developed by the shock of blasting in quarries, or road-cuttings ; it is very infrequent in natural exposures. The curvature is sometimes so acute that the rocks are divided up into spherical or ellipsoidal blocks which have occasionally been mistaken for pillow structures.

Felsites and other acid rocks are often intersected by such closely-spaced, irregular, joint planes that, on weathering, they break into small, sharply-angular fragments from which it is difficult to obtain a hand specimen of the normal size. One set of closely-spaced joint planes may be so well developed as to produce a *platy* fracture, which is often prominent in such rocks as phonolite and mugearite. This platy parting is sometimes dependent upon an obscure flow structure, as, for example, in the mugearites of the Scottish Carboniferous.

Jointing is sometimes caused by tensile stress consequent upon contraction due to cooling. In other cases tectonic causes, producing tensional, compressional, or torsional stresses, originate the fracturing.

COLUMNAR AND PRISMATIC STRUCTURES—With uniform cooling and contraction in a homogeneous magma the parting-planes tend to take on a regular columnar or prismatic form, characterised by the development of four-, five-, or six-sided prisms, which may be intersected by cross-joints (Fig. 23 A). This is the phenomenon illustrated mainly by basaltic rocks, as at the Giant's Causeway and Staffa, but occasionally by other rock types. The columns develop perpendicularly to the cooling surfaces so that in a sill or lava flow they stand vertically, whilst in a dyke they are more or less horizontal.

Columnar structure is due to the development of centres of contraction at equally-spaced intervals on the cooling surfaces. The lines joining these centres are the directions of greatest tensile stress, and when the rigidity of the rock is overcome, cracks will appear perpendicularly to these lines. As seen in Fig. 23 B, these cracks will intersect so as to enclose a hexagonal area. As they extend downwards or upwards or laterally from the cooling surfaces, they produce the well-known prismatic columns. The columns extending from

[1] Tyrrell and Martin, *Trans. Geol. Soc., Glasgow*, xiii, pt. 2, 1908 p. 243 (Pl. XIII, Fig. 2).

opposite cooling surfaces often join along a well-marked median plane.[1] The four-, five-, or even seven-sided prisms result from irregularities in the spacing of the contraction centres.

Sosman has shown that it is possible for prismatic structures to develop in igneous rocks by convectional circulation whilst the magma is still in the liquid condition.[2] By this means a state of subdivision into irregular cells of from four

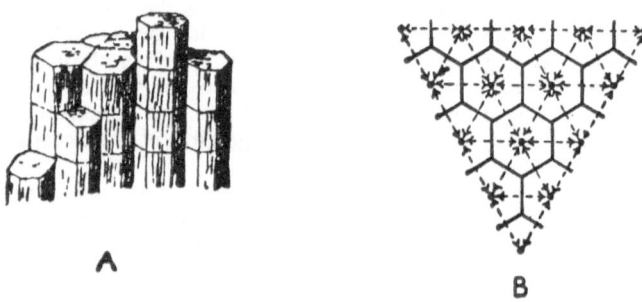

A

B

FIG. 23.—COLUMNAR JOINTING.

A. Columnar jointing in basalt.
B. Explanation of columnar jointing as due to cooling tensions developed equally about uniformly-spaced points. The arrows indicate the directions of the forces acting about each point.

to seven sides is produced, which may leave a record in the solid rock by causing magmatic segregation in the cell walls and axes, or by originating regularly-spaced centres of crystallisation.   This process, however, has not yet been fully established as a cause of prismatic structure in igneous rocks, although certain columnar basalts in Auvergne have been attributed to it by French investigators.

RIFT AND GRAIN—In quarrying granite advantage is taken of the mural jointing to procure large blocks; but in dressing the blocks down to smaller dimensions quarrymen make use of the *rift* and *grain*, which are directions of comparatively easy splitting at right angles to one another, one

[1] *Mull Memoir*, 1924, p. 108.
[2] *Journ. Geol.*, 24, 1916, p. 219.

horizontal and the other vertical. In the direction at right angles to both the rift and the grain, the granite breaks with a rough irregular fracture which is called the " hard way," the " tough way," or the " head." Rift is the direction of easiest fracture, the " cleaving way " in Cornwall, and the " reed " in Scotland. Along the grain, the " quartering way " (Cornwall), or the " hem " (Scotland), fracture is less perfect. The perfection of rift and grain determines the ease with which kerb stones and setts can be made.

According to T. N. Dale [1] rift and grain structures consist

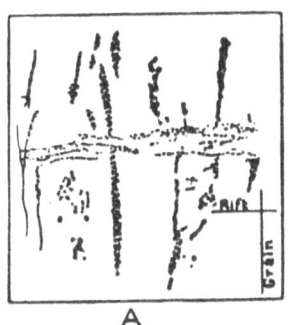
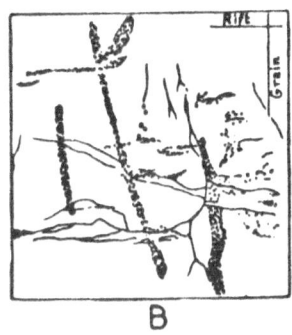

FIG. 24.—RIFT AND GRAIN.

Sheets of cavities and cracks in quartz grains from granite, Conway, New Hampshire, approximately parallel to rift and grain directions. (A) Enlarged about 60 diameters ; (B) Enlarged about 20 diameters. After T. N. Dale, *Commercial Granites of New England : Bull. 738, U.S. Geol. Surv.*, 1923.

of minute cracks from ·09 to 1·3 mm. apart, crossing the quartz particles and extending into the felspars. These cracks determine the rough fissility in two rectangular directions, of which the usually horizontal rift is the more pronounced. These fractures coincide with, or are parallel to, bands of fluidal cavities in the quartz grains. These sheets cross one another at right angles, and those in the grain direction are the less abundant (Fig. 24). Both sets are independent of sheet or flow structures in the granite, but

[1] *Op. cit.*, pp. 15-26

the rift is frequently parallel to mica flakes and to the long
axes of felspar phenocrysts when these are present. Rift
and grain are not pronounced in all granites. Either or
both may be ill-developed or absent. These structures may
be present in other igneous rocks, but usually to much less
extent than in granite.

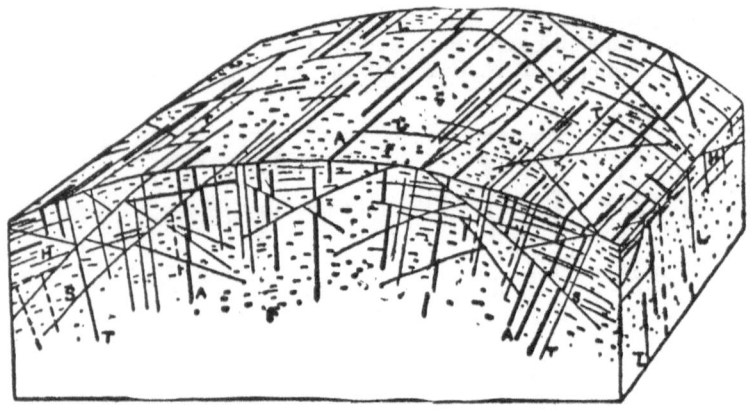

FIG. 25.—STEREOGRAM OF RIFT, GRAIN, AND JOINTING IN GRANITE.
Diagram of a domed sheet of granite. L, longitudinal jointing and
   grain direction; H, horizontal jointing and rift direction; T, transversal or cross jointing; F, direction of oriented, fluxionally-arranged
   mineral constituents; A, Aplite dykes in widened transversal and
   horizontal fissures; S, surfaces of stretching and movement which
   fault the older joints and dykes. Adapted from Cloos, *Das Batholithenproblem*, 1923.

Note to Fig. 25.—It will be noticed that Cloos introduces a third
direction of parting and jointing (cross-fracture) due to the bending and
stretching of the granite as an arched sheet. Aplite and other dykes
find their way mainly into the open fissures thus produced.

The production of rift and grain is dependent on the
crustal stresses which were in operation at the time of consolidation of the granite. The sheets of fluidal cavities were
formed when quartz, the final constituent, was crystallising.
Under the same stress the mica plates became aligned in
directions roughly parallel to the rift. Then as crustal
stresses continued to operate after the rock had solidified,

the rift and grain cracks were produced, following the planes of weakness developed by the cavities.

Professor H. Cloos and his school believe that not only the rift and grain, but that all the directional structures in granite, including joints and dykes, are related to the crustal stresses which prevailed during and after the consolidation of the granite (Fig. 25).   Hence the study of these directions may provide valuable information as to the nature and direction of the earth forces which were concerned in their production, and which were operative in the mountain-building coincident with the intrusion of the granite.[1]

[1] H. Cloos, *Tektonik und Magma*, Band 1. *Abh. d. Pr. Geol. Landes-anst*, N.F. 89, Berlin, 1922. For explanations of Cloos's views in English see G. W. Tyrrell, *Science Progress*, Oct. 1925, pp. 211-3; R. Balk, *Bull. Geol. Soc. Amer.*, 36, 1925, pp. 679-96. Cloos's methods have been applied to the study of the Dartmoor granite by A. Brammall, *Proc. Geol. Assoc.*, xxxvii, 1926, pp. 251-77; and to the Scilly Isles granite by C. W. Osman, *Quart. Journ. Geol. Soc.*, lxxxiv, 1928, pp. 258-92.

CHAPTER III

COMPOSITION AND CONSTITUTION OF MAGMAS

THE MAGMA—The evidence upon which it is believed that a large number of the rocks of the earth's crust are of igneous origin has been briefly detailed on page 11. The original molten rock matter is conveniently termed *magma*. Etymologically magma means a thick, pasty, porridge-like mass, and it is, therefore, entirely appropriate as applied to viscous molten material carrying crystals in process of formation. Present-day lavas, especially the molten lakes such as that of Halemaumau in Hawaii, are the most accessible and easily-studied representatives of magmas.

COMPOSITION OF MAGMAS—On page 6 the average composition of the 10-mile crust is given; and as igneous rocks are estimated to comprise 95 per cent. of this bulk, the average composition of igneous rocks differs only slightly from the figures for the crust. Table II, column 1, shows the composition of the average igneous rock in percentages of elements, and column 2 in the form of oxides, as computed by Clarke and Washington from over 5000 analyses.[1]

Practically all the known elements have been met with in igneous rocks, but only nine, namely, oxygen, silicon, aluminium, iron, calcium, sodium, potassium, magnesium, and titanium, can be regarded as at all common, and make up between them 99·25 per cent. of the whole. Nevertheless, some of the rarer elements, especially volatile ones such as hydrogen, fluorine, chlorine, and sulphur, as well as their volatile compounds, are of the greatest importance in the phenomena of igneous rocks (see p. 159). Because of the tendency of the more volatile constituents to escape before

---

[1] "Composition of the Earth's Crust," *Prof. Paper* 127, *U.S. Geol Surv.*, 1924, pp. 6-16.

TABLE II

1		2	
Oxygen	46·59	SiO_2	59·12
Silicon	27·72	Al_2O_3	15·34
Aluminium	8·13	Fe_2O_3	3·08
Iron	5·01	FeO	3·80
Calcium	3·63	MgO	3·49
Sodium	2·85	CaO	5·08
Potassium	2·60	Na_2O	3·84
Magnesium	2·09	K_2O	3·13
Titanium	·63	H_2O	1·15
Phosphorus	·13	CO_2	·102
Hydrogen	·13	TiO_2	1·050
Manganese	·10	ZrO_2	·039
Sulphur	·052	P_2O_5	·299
Barium	·050	Cl	·048
Chlorine	·048	F	·030
Chromium	·037	S	·052
Carbon	·032	$(Ce,Y)_2O_3$	·020
Fluorine	·030	Cr_2O_3	·055
Zirconium	·026	V_2O_3	·026
Nickel	·020	MnO	·124
Strontium	·019	NiO	·025
Vanadium	·017	BaO	·055
Cerium and Yttrium	·015	SrO	·022
Copper	·010	Rest	·023
Remainder of elements	·034		
	100·000		100·000

and during solidification, analyses of the igneous rocks do not fairly represent the composition of magmas. Observations at volcanic vents, and the collection of gases from molten lavas, have shown conclusively that volatile constituents are present in magma in very much greater amounts than are recorded in igneous rock analyses. Water alone is estimated to make up about 4 per cent. of the basaltic magma of Kilauea.

THE PYROGENETIC MINERALS—The minerals formed from igneous magmas are termed *pyrogenetic* (i.e. formed by fire). Since oxygen and silicon are the most abundant elements in magmas silicates and silica form the chief igneous minerals. A few other oxides occur besides silica ; but other compounds are present only in insignificant amounts, although they are important in a petrogenetic sense.

Most igneous rock minerals are, therefore, silicates. Silicon is capable of forming several acids, of which the following are the most important from the petrographical point of view :—

Orthosilicic acid, H_4SiO_4 or $2H_2O, SiO_2$.   Ex. olivine, $2 (Mg, Fe)O, SiO_2$.

Metasilicic acid, $H_4Si_2O_6$ or $2H_2O, 2SiO_2$.   Ex. enstatite, $(Mg, Fe)O, SiO_2$.

Polysilicic acid, $H_4Si_3O_8$ or $2H_2O, 3SiO_2$.   Ex. orthoclase, $K_2O, Al_2O_3, 6SiO_2$.

Hence pyrogenetic silicates fall into the three groups of orthosilicates, metasilicates, and polysilicates, distinguishable, as are the acids, by the oxygen ratios of bases to silica being respectively $1 : 1$, $1 : 2$, and $1 : 3$. The olivine group, varying from forsterite, $2MgO,SiO_2$, through olivine, $2(MgFe)O,SiO_2$, to fayalite, $2FeO,SiO_2$, forms perhaps the most important rock-forming orthosilicate. Nepheline, $Na_2O,Al_2O_3,2SiO_2$, and anorthite, $CaO,Al_2O_3,2SiO_2$, are also orthosilicates. The pyroxenes and amphiboles are common examples of the metasilicate group, thus, for example, diopside, $CaO,(MgFe)O,2SiO_2$ and hypersthene, $(MgFe)O,SiO_2$. Leucite, $K_2O,Al_2O_3,4SiO_2$, is also a meta-silicate. Pyrogenetic polysilicates are best represented by orthoclase, $K_2O,Al_2O_3,6SiO_2$, and albite, $Na_2O,Al_2O_3,6SiO_2$.

Potassium and sodium form the most active bases present in igneous magmas ; calcium is less active ; magnesium and iron are relatively the weakest. Hence silica is taken up to the fullest extent chiefly by the alkali metals, which, therefore, usually form polysilicates, the highest possible amount of silica being bound up with them in orthoclase and albite. It is to be noted that aluminium only enters into combination with silica along with a base-forming element, usually in equal molecular amounts. Calcium tends to form meta-silicates ; magnesium and iron both metasilicates and ortho-silicates. Since iron has the weakest affinity for silica in this series, it is often left out of combination if there is a deficiency of silica, and appears as oxide (magnetite). If there is a considerable deficiency of silica in the magma, potassium and sodium may not be able to combine with sufficient silica to form orthoclase and albite respectively, and the lower silicates leucite and nepheline will be formed instead. As potassium has the stronger affinity for silica of the two, the albite molecule will first experience the desilication, and nepheline will be formed in preference to leucite.

Similarly, in magmas with relatively large amounts of magnesium and iron, pyroxenes are formed when there is an adequate amount of silica, but olivine when silica is deficient. Rock-forming silicates may thus be arranged in two groups of low and high silication respectively, such that the following equation holds :—

Mineral of low silication $+$ silica $\rightleftarrows$ mineral of high silication ; and the reaction is reversible, i.e. it may run in either direction.

Minerals of Low Silication.	Minerals of High Silication.
Leucite.	Orthoclase.
Nepheline	Albite.
Analcite.	Anorthoclase.
Olivine.	Orthorhombic pyroxenes.
	Augite.
Biotite.	Aegirine.
	Hornblende

Some silicates, especially analcite, the micas, and hornblende, contain combined water, which appears to be essential to their constitution. Halogen elements, especially fluorine, appear to be active in the formation of certain micas.

Any excess of silica which may be left over after the bases are fully satisfied crystallises out as quartz. Consideration of the above equation will show that quartz cannot co-exist with minerals of low silication (with the exception of biotite) save under the most unusual conditions. Hence quartz is never found in the same igneous rock as leucite, nepheline, analcite, and but rarely with olivine, a very convenient fact when the classification of igneous rocks comes to be considered (see p. 104).

The minor constituents of igneous magmas crystallise out as accessory constituents which are small in bulk, but are frequently of very wide distribution. Ilmenite $(Fe,Ti)_2O_3$; apatite, $Ca_3(PO_4)_2$, CaF_2; sphene, CaO,TiO_2, SiO_2; and zircon, ZrO_2,SiO_2; are the most important of these. In addition a large number of uncommon minerals containing rare elements have been found in igneous rocks.

Some minerals appear, however, to be never of pyrogenetic origin. Of these, the typical metamorphic minerals, such as the silicates of aluminium, andalusite, kyanite, and silli-manite, along with cordierite, staurolite, etc., are the most prominent. Pure alumina, however, crystallised as corun-dum, may occur in igneous rocks under exceptional circum-stances.

THE PHYSICO-CHEMICAL CONSTITUTION OF MAGMAS—Con-sideration of the chemical composition of minerals which actually crystallise from magmas leads to certain conclusions as to the molecules which are present in the molten state. Niggli [1] enumerates the following fourteen molecules as the principal constituents of igneous magmas :—

1. $[SiO_4]^{Al}_K$ . . . Kaliophilite.

2. $[SiO_4]^{Al}_{Na}$ . . . Nepheline.

3. $[SiO_4]^{Al_2}_{Ca}$ . . . Tschermak's molecule.

4. $[SiO_4]^{Al}_H$ . . . Mica.

5. $[SiO_4]Na_4$ . . . Sodium silicate.
6. $[SiO_4]K_4$ . . . Potassium silicate.
7. $[SiO_4]Ca_2$ . . . Calcium orthosilicate.
8. $[SiO_4]Fe_2$ . . . Fayalite.
9. $[SiO_4]Mg_2$ . . . Forsterite.
10. $[SiO_2]$ . . . Quartz.
11. $[H_2O]$ . . . Water.
12. $[O_2]$ . . . Oxygen.
13. $[Fe_2O_4]Fe$ . . . Magnetite.
14. $[Al_2O_4]Mg$ . . . Spinel.

Numerous other molecules are undoubtedly present, but only in very small amounts.

It will be noted that all the silicates are of orthosilicate type ; the only oxides are those of hydrogen, silicon, iron ;

[1] *Lehrbuch der Mineralogie*, 1921, p. 478. The notation used by Niggli is retained in the ensuing formulæ.

and the only non-silicate aluminate is that of magnesium (spinel). The essential pyrogenetic minerals are formed either by the individual separation of the above constituents, or by additive reactions between them. The silica molecule is especially prominent in these reactions. Thus the addition of two molecules of silica to kaliophilite and nepheline leads to the formation of orthoclase and albite respectively; and all possible varieties of the common felspars may be formed by combinations between kaliophilite, nepheline, Tschermak's molecule, and silica.

$$[SiO_4]^{Al}_K + 2SiO_2 = [SiO_4, SiO_2, SiO_2]^{Al}_K$$
(orthoclase)
$$[SiO_4]^{Al}_{Na} + 2SiO_2 = [SiO_4, SiO_2, SiO_2]^{Al}_{Na}$$
(albite)
$$[SiO_4]^{Al_2}_{Ca} + SiO_2 = [SiO_4, SiO_2]^{Al_2}_{Ca}$$
(anorthite)

Olivine is formed by the combination of the fayalite and forsterite molecules. The complex pyroxenes and amphiboles are produced by combinations in various proportions of calcium orthosilicate, fayalite, forsterite, quartz, Tschermak's molecule, and spinel. Biotite comes into being by combination between kaliophilite, mica, fayalite, and forsterite.

The physico-chemical condition of magmas must be one of mutual solution of all constituents. Bunsen, in 1861, first suggested that magmas were comparable to complex solutions of salts in water. Magmas, however, are solutions of peculiar composition. The principal molecules are silicates, difficultly-volatile and refractory substances of high melting-point and low vapour pressure; which are associated with molecules of quite different physico-chemical characters, namely, gases and vapours, such as water, sulphuretted hydrogen, hydrofluoric acid, hydrochloric acid, carbon monoxide, carbon dioxide, sulphur dioxide, hydrogen, nitrogen, and oxygen. These are volatile substances of high vapour pressure, which tend to diminish the viscosity of the magma, lower the freezing-points of the silicates, and thus facilitate crystallisation. The gaseous constituents also form volatile compounds with some of the other molecules present, and with them migrate in the direction of lessened pressure, thus producing partial separations within the magma. This contrast in physico-chemical characters between the two principal groups of magmatic constituents (*fixed* and *fugitive*) has most important consequences, as it is instrumental in producing many of the variations in igneous rocks, and also many of the concentrations of useful metallic substances which are known as ore deposits (p. 160).

The relative numbers of the different molecules which may be present in any given magma are regulated by the external conditions, of which temperature and pressure are the most important. Put in another way, there is *equilibrium* between the constituents which is disturbed when the external conditions vary. Equilibrium is restored by interchanges between the molecules. Various reactions may thus be started which may result in one or more constituents reaching such a concentration that crystallisation begins. Further, it is easy to see that a magma of given composition may give rise to more than one set of minerals under different conditions. Thus a mixture of orthoclase and biotite in certain proportions has the same bulk composition as a mixture of leucite and olivine in certain proportions. The equivalence is illustrated by the following equation :—

$$2\ KAlSi_3O_8 + K_2Mg_2Al_2Si_3O_{12} = 4KAlSi_2O_6 + Mg_2SiO_4$$

orthoclase   +   biotite   leucite + olivine
   (Mica-syenite)   (Leucite-basalt)

Orthoclase and biotite is the mineral association which will crystallise from a magma of the above composition at comparatively low temperature and high pressure ; while the combination leucite and olivine is stable at high temperature and low pressure. Hence the former rock is a plutonic (deep-seated), and the latter a volcanic product.

Another example is afforded by the following equation :—

$$CaMgSi_2O_6, MgAl_2SiO_6 = Mg_2SiO_4 + CaAl_2Si_2O_8$$

Augite   = Olivine + Anorthite

Thus a rock composed entirely of augite (augitite, a lava) may be of the same chemical composition as one composed of olivine and anorthite (allivalite, a plutonic rock).

The phenomenon of rocks of different mineral composition arising from similar magmas under different conditions of crystallisation is termed *heteromorphism* by Lacroix,[1] and the products are said to be *heteromorphs* of each other.   Thus leucite-basalt is a heteromorph of mica-syenite, and allivalite of augitite.   Further discussion of the constitution and composition of magmas will be found in succeeding chapters.

PRIMARY MAGMAS—The average igneous rock, the composition of which is given in Table II, is an abstraction, there utilised for the discussion of magmatic composition in general. It only incidentally corresponds with the composition of certain actual igneous rocks, and those by no means the most abundant.   It is necessary, then, to enquire as to the nature and composition of the magma or magmas from which igneous rocks have actually been derived.   It is out of the question that the source of every variety of igneous rock was a magma of corresponding composition.   Field and laboratory experience shows that igneous rocks are products of differentiation from a limited number of magmas ; and if the origin of these sub-magmas be traced far enough back, it is believed that the assumption of two primary magmas at most, will suffice to account for the known variations of igneous rock composition.

This conclusion is based on the frequency with which the various rock types occur.   It is a matter of common observation that granite (or rather granodiorite, see p. 113), and basalt, are by far the most abundant igneous rock types. Granites, granodiorites, or other granitoid rocks, form the greatest batholiths which occur in the earth's crust (see p. 29) ; basalts, on the other hand, form enormous plateaux, which, in certain cases, cover areas of the order of hundreds of thousands of square miles.   Professor Daly has tested this observation statistically by measuring the areas covered by the principal igneous rock types in North America.[2]   On this basis he comes to the conclusion that granites and granodiorites together have more than twenty times the total area of all other intrusive rocks combined ; and that

[1] *Comptes Rendus*, Paris, tom. 165, 1917, p. 486 ; tom. 170, 1920, p. 23.
[2] *Igneous Rocks and Their Origin*, 1914, Chap. III.

basalt has probably at least five times the total volume of all other extrusives combined. Amongst the other volcanic types pyroxene-andesite is the most abundant next to basalt, and of plutonic rocks gabbro is the most abundant next to the granitic types. But the andesites in general are the effusive equivalents of granodiorite, as can be shown by comparing the average compositions of these types; [1] and the gabbros are the plutonic equivalents of the basalts. Hence by far the most abundant magmatic types appearing in the earth's crust are the granodiorite-andesite magma, and the basalt-gabbro magma. The chemical composition of average granodiorite and of average plateau basalt is shown in Table III below.

TABLE III.

	1	2 [2]
SiO_2	65·1	49·3
Al_2O_3	15·8	14·1
Fe_2O_3	1·6	3·4
FeO	2·7	9·9
MgO	2·2	6·4
CaO	4·7	9·7
Na_2O	3·8	2·9
K_2O	2·3	1·0
H_2O	1·1	—
TiO_2	·5	2·6
MnO	·1	·2
P_2O_5	·1	·5
	100·0	100·0

1. Granodiorite, mean of 12 analyses. Daly, *Igneous Rocks and Their Origin*, 1914, p. 25.

2. Plateau-basalt, mean of 50 analyses computed by R. A. Daly ("Geology of Ascension Island," *Proc. Amer. Acad Arts and Sciences*, 60, 1925, p. 73), from H. S. Washington's analyses (*Bull. Geol. Soc. Amer.*, 33, 1922, p. 797).

The statistical study of igneous rock analyses by Dr. W. A. Richardson and Mr. G. Sneesby has led to similar conclusions.[3] If a curve be constructed showing the number of analyses in

[1] *Igneous Rocks and Their Origin*, 1914, Chap. III. Cf. anal. 46 (p. 26) with anal. 41 (p. 25).
[2] Computed as water-free.
[3] *Min. Mag.*, xix, 1922, pp. 303-13.

which the silica percentages fall between certain limits (in this case a 1 per cent. interval was used), two peaks appear, one at 52·5 per cent. and the other at 73 per cent. (Fig. 26). In other words, these silica percentages are the most frequent in igneous rocks.   If similar curves be constructed for the other oxides, it can be shown that the two most frequent igneous rocks so far sampled for analysis have compositions representing granite and acid basalt or dolerite respectively. The silica percentages of both granite and basalt obtained in this way are somewhat too high to correspond exactly with

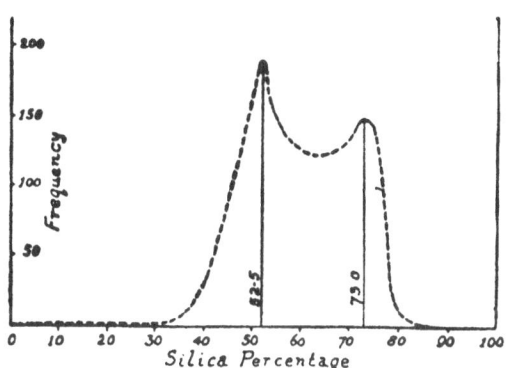

FIG. 26.—FREQUENCY DISTRIBUTION OF IGNEOUS ROCKS.
Curve of frequency based on silica percentages.   See text, p. 54.   From
W. A. Richardson and G. Sneesby

Daly's results.   This is probably because the great granodiorite batholiths and the extensive plateau basalt floods have so far been inadequately sampled for analysis.   When these masses are fully represented in collections of igneous rock analyses, the frequency curve will probably show maxima at points somewhat below those yielded by present data.

It is worth while noting the harmony between these results and the composition of the crust as deduced from quite different data (Chap. I).   From geodetic observations, such as the determination of the force of gravity, and from considerations based on the doctrine of isostasy, it is believed that the continents consist mainly of rock of granitic composition (*sial*) resting upon, and passing into, a substratum (*sima*), the upper part of which is of basaltic composition. This basaltic substratum is believed to form the floor of the oceans without the intervention of acid rocks. Hence the frequency of granitic and basaltic igneous rocks is in accordance with this view of the composition and structure of the earth's crust.

The conception of two major primary magmas, basic and acid, which, by various degrees of admixture, form many groups of igneous rocks, notably those of Iceland, was first formulated by Bunsen in the middle of last century. From considerations based on the petrology of the tholeiite dykes (see p. 129) of the North of England, A. Holmes and H. F. Harwood have revived Bunsen's hypothesis as an explanation of the origin of these rocks.[1] It is shown that their variations cannot be due to ordinary crystallisation-differentiation (see p. 153). The rocks behave as basaltic magmas which, in some cases, have been merely diluted with a quartz—alkali-felspar mixture. It is therefore concluded that tholeiites may possibly be the products of the partial mixing of basaltic and granitic magmas generated at their respective sima and sial levels in the crust.

[1] *Min. Mag.*, xxii, 1929, pp. 1-52.

CHAPTER IV

THE FORMATION OF IGNEOUS ROCKS

GLASS AND CRYSTALS—The solid matter of which igneous rocks are composed consists either of non-crystalline material (glass), of crystals, or of both crystals and glass. Magmas become igneous rocks by solidification without crystallisation (formation of glass), or by crystallisation, with the loss of much of their volatile material. The difference between crystalline and glassy matter is analogous to the difference between a disciplined battalion and a scattered mob. The molecules or ultimate particles of a crystal are arranged in a definite manner; they are, as it were, neatly packed. In a glass, however, the molecules have settled down into a stable arrangement but without any recognisable pattern. A glass must be regarded as an amorphous solid, not as a super-cooled liquid possessing high rigidity.

A glass is less dense than the corresponding crystalline substance, because its molecules are not so neatly and closely arranged as those of the crystal. Thus a cubic inch of crystalline silica (quartz) weighs 650 grains, but a cubic inch of glassy silica (melted quartz) only 560 grains. It follows that under great pressure crystalline matter will form more readily than glass, because the crystalline state represents the greatest economy of space under the prevalent conditions. Conversely, glass is more readily formed from magmas under conditions of low pressure. Magmas which have crystallised under a heavy load of superincumbent rock at considerable depths in the crust therefore form completely crystalline rocks; whereas in those which are erupted to the surface glassy matter is frequently present.

CRYSTALLISATION OF A UNICOMPONENT MAGMA—Magmas usually consist of many components, a few of which greatly preponderate over the others. Hence most igneous rocks are multicomponent; they consist of three, four, or five principal mineral constituents, with a number of minor ones. Igneous rocks of two principal components only, are known, but are uncommon; and unquestionably unicomponent rocks are extremely rare.

In order to study the solidification of magmas we shall take them in order of increasing complexity, starting with a magma containing only one constituent. This comes to practically the same thing as the consideration of the crystallisation of a single pure mineral from a melt of its own composition.

A rock-forming mineral which has been investigated in this way is augite.[1]   The crystallisation of augite is indicated by the diagram, Fig. 27, in which the abscissa represents the

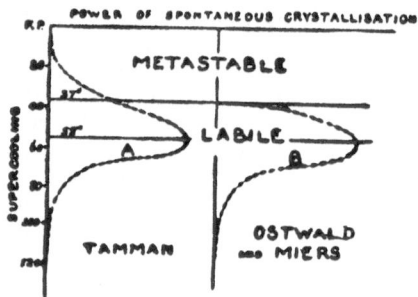

FIG. 27.—CURVES ILLUSTRATING THE SUPERCOOLING OF AUGITE.
See text, p. 58.   From Harker, *Natural History of Igneous Rocks*, 1909.

power of spontaneous crystallisation, measured by the number of crystals initiated in unit volume in unit time, and the ordinate represents temperature below the freezing-point.

According to Tamman the curve A represents the mode of crystallisation.   Crystallisation begins at the freezing-point, but the rate is at first extremely slow.   Between $30°$ and $55°$ below the freezing-point the number of centres of crystallisation increases enormously, reaching a maximum at $55°$, and then declining again as rapidly, until crystallisation ceases at a temperature $120°$ below freezing-point.

This illustrates the phenomenon of *super-cooling* (*under-*

[1] Doelter, *Phys. Chem Min.*, 1905, pp. 111-12.

cooling) or *supersaturation ;* i.e. the main onset of crystallisation does not take place at the freezing-point, but at some distance below it. The temperature region in which the generation of crystals is slow is called the *metastable* region ; that in which the rate of crystallisation is rapid is the *labile* region. According to Ostwald and Miers, however, the curve B better represents the course of crystallisation. Their views were developed from the study of aqueous solutions of salts, and are here transferred to a rock-forming mineral. They believe that no crystallisation occurs at the freezing-point unless the solution is inoculated with a crystal of the dissolved substance; but if inoculation is prevented crystals suddenly appear in great numbers when the labile region is reached. The experience of petrologists is more in accord with the views of Tamman and Doelter than with those of Ostwald and Miers.

GRAIN OF IGNEOUS ROCKS.—The above leads to an explanation of the variations of grain size in igneous rocks. If the cooling of the magma is prolonged it spends a protracted time within the metastable region of crystallisation ; only a few centres of crystallisation are initiated, and these crystals grow to a large size. If cooling is very slow the magma may be completely crystallised before the labile region is reached. Hence slow cooling leads to coarse grain. On the other hand, if the cooling is rapid, the metastable region is quickly passed over, and the bulk of crystallisation takes place in the labile region. A great number of crystals are initiated, and the resulting rock is fine grained. Crystallisation liberates heat, which still further prolongs the period in which the magma is within the crystallising range of temperature.

In magmas composed of several constituents, the special properties of each constituent are modified by the presence of the others, the power of spontaneous crystallisation amongst them. According to Tamman, the velocity of crystallisation of any constituent is increased by the presence of other substances. Hence, other things being equal, a multicomponent igneous rock should be of finer grain than one of simpler composition. Of gabbro and anorthosite, for example, which are rocks of comparable chemical composition, the more complex type, gabbro, is usually of finer grain than anorthosite, which is mainly composed of labradorite.

Grain size is also affected by at least two other considerations, namely, the molecular concentration of the substances present, and the viscosity of the magma. Components present in very small amounts, as, for example, the molecules of such minerals as apatite, zircon, magnetite, etc., tend to form small crystals because the available material for their growth is quickly exhausted.

The growth of a crystal in a magma implies diffusion of its molecules towards the centre of crystallisation. If the magma is viscous it opposes a strong passive resistance to diffusion. Hence crystallisation is retarded by viscosity, and, other things being equal, the more viscous magmas give rise to the finer grained rocks. Rhyolitic and felsitic magmas are much more viscous than basaltic magmas, as can be inferred from their respective modes of eruption and flow (p. 14). Corresponding with this, rhyolites and felsites are much finer in grain than are basalts erupted under similar conditions.

Any circumstance or condition which tends to reduce the viscosity of a magma promotes the growth of crystals. The presence of water and other gases and fluids in magmatic solution greatly increases the liquidity, and crystals grow to larger sizes in magmas rich in these constituents than in " dry " magmas. The formation of pegmatites (p. 162), with their giant crystals, is directly dependent on great concentration of volatile constituents in magmatic residues.

FORMATION OF GLASS [1]—Reverting to the curve, Fig. 27 A, it is easy to see that if the cooling were very rapid, a magma might pass through both metastable and labile regions with little or no crystallisation. As the possibility of crystallisation ceases at some temperature not far below the freezing-point, 120° below in the case of augite, the liquid solidifies as a glass under these conditions. Its molecular condition certainly changes, and it suffers an enormous increase of viscosity. According to recent work a glass must not, however, be regarded as an excessively viscous super-cooled liquid. If crystallisation takes place to some extent before the temperature drops below the lower limit, the unexhausted residue

[1] L. Hawkes, *Geol. Mag.*, lxvii, 1930, pp. 17-24.

of liquid solidifies as a glass which forms a groundmass to the crystals.

Fine grain and the presence of glass in an igneous rock are therefore indications of rapid cooling. Hence these features are common on the margins of intrusive igneous masses, where magma has come into contact with cold rock. They are especially common in lavas which have been rapidly cooled by contact with the atmosphere or with water. As viscosity retards crystallisation, viscous magmas, such as those of rhyolites, more often form glassy rocks than the more mobile magmas, such as those which give rise to basalts. Hence obsidian and pitchstone, glassy forms of acid magmas, are much more abundant than tachylyte, the glassy form of basalt.

CRYSTALLISATION OF BINARY MAGMAS—The crystallisation of bi-component magmas can be easily understood if the fundamental principle is kept in mind that the specific properties of each constituent are modified in the presence of the other. Most important is the fact that the freezing-points are lowered. This fact is illustrated by the phenomena of freezing mixtures, such as ice and salt. Water freezes at $0°$ C.; molten salt solidifies at about $800°$ C.; but a mixture of ice and salt in certain proportions freezes at $-22°$ C. The lavas of Vesuvius furnish a petrographical example of this property. Augite is frequently found enclosed in the leucite of these rocks, and is therefore of earlier crystallisation. The freezing-point of Vesuvian augite may be taken as $1220°$ C. Hence the leucite must have crystallised at some temperature lower than $1220°$; but as leucite by itself freezes at $1420°$, the presence of augite (and other constituents) has lowered its freezing-point by at least $200°$.

The crystallisation of a magma consisting of two independent constituents may be illustrated by the use of the temperature-composition diagram (Fig. 28), in which the abscissa represents composition, and the ordinate temperature. As an example a magma consisting of a mixture of the minerals A (freezing-point $Ta°$) and B (freezing-point $Tb°$) may be taken. A magma consisting entirely of A would crystallise at $Ta°$, and its freezing-point is represented by the point P on the diagram. Similarly, a magma consisting entirely of B would crystallise at Tb°, the freezing-point being represented by the point R. The addition of 10 per cent. of B to the A magma, producing magma of the composition $A_{90}B_{10}$, causes a certain lowering of the freezing-point which is now represented by the point P_1. The addition of 20 per cent. of B causes a further lowering of the freezing-point to P_2, and so on. A similar lowering of freezing-point is produced in a B magma by the addition

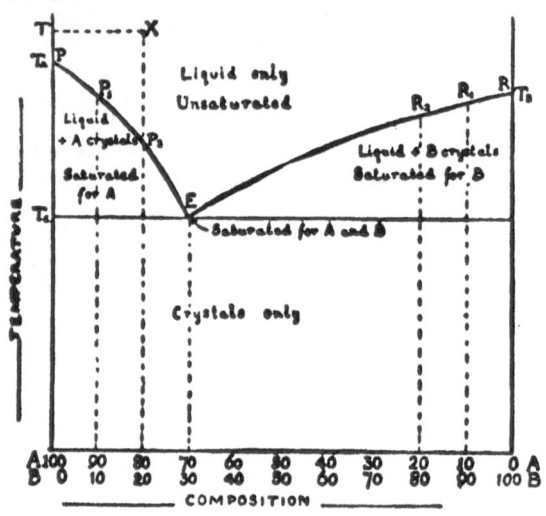

FIG. 28.—TEMPERATURE-COMPOSITION DIAGRAM
Illustrating the formation of eutectics, etc. (text, p. 63).

of A, and may be represented by the points R_1, R_2, etc. By joining up these points we obtain the curves PE and RE, meeting in E, which represent the freezing-points of any possible mixture of A and B.

Any point on the diagram, such as X, indicates a definite condition of the magma in regard to temperature and composition. Any change of condition may be indicated by movement of a point. Thus horizontal movement means change of composition at constant temperature. Vertical movement means change of temperature at constant composition. Diagonal movement means change both of temperature and composition.

The curves PE and RE may also be regarded as *saturation* curves. Magmas represented by points above the curves are unsaturated. Points on the curve PE indicate magmas saturated with A; and points on the curve RE magmas saturated with B. The point E represents a magma saturated with both A and B.

Let us follow the crystallisation of a magma of composition $A_{60}B_{30}$ at a temperature, say, of T°. This will be represented by the point X on the diagram. As the temperature falls without change of composition of the magma, the indicator point X moves vertically downwards until the point P_2 is reached. At P_2 the magma is saturated with A, which forthwith begins to crystallise. A is thereby abstracted from the magma which, therefore, becomes enriched in the B component, with a consequent fall in the freezing-point along the curve PE. Hence the indicator point moves down the curve, the diagonal movement indicating both the lowering of temperature and the change in composition of the magma, until the point E is reached. At this point the magma becomes saturated with B, and A and B separate out together at constant temperature until the magma is exhausted. This occurs at temperature Te, and magmatic composition $A_{70}B_{30}$.

In magmas with more than 30 per cent. B, B will separate out first, and the indicator point will move down the curve RE until the point E is reached, when A also begins to crystallise.

Thus the process of crystallisation in a binary magma takes place in three stages: (1) cooling of magma prior to crystallisation; (2) crystallisation of one constituent which is in excess of certain standard proportions ($A_{70}B_{30}$ in the above case) with falling temperature; (3) simultaneous crystallisation of both constituents at constant temperature.

EUTECTICS—The constant proportion in which the two constituents simultaneously crystallise (point E in Fig. 28) is called the *eutectic*. The intergrowth of the two minerals frequently results in a peculiar *graphic* texture, in which the constituent present in smaller quantity is embedded as apparently disconnected patches in the other, with shapes often recalling cuneiform writing (Fig. 29), and possessing the same optical orientation over a considerable area. The converse does not hold; not all graphic textures are of eutectic origin.

The mineral which is present in excess of the eutectic proportions often forms somewhat larger crystals embedded in a ground mass of constant (eutectic) composition, and of finer grain.

Vogt has determined the following eutectic proportions for certain pairs of rock-forming minerals :—

> Orthoclase : Quartz : : 72·5 : 27·5.
> Orthoclase : Albite : : 42 : 58.
> Anorthite : Olivine : : 70 : 30.
> Diopside : Enstatite : : 45 : 55.

The eutectic intergrowth of orthoclase and quartz produces the well-known rock *graphic granite* (Fig. 29); when on a microscopic scale, the intergrowth is called *micropegmatite*. *Perthite* and *microperthite* are the corresponding terms for coarse and fine intergrowths of orthoclase and albite.

With magmas of three independent constituents crystallisation is more complex, but follows a similar course to that of bi-component solutions. The constituent which first saturates the magma crystallises first with falling temperature. Then, when the saturation point of the second constituent is reached, a binary eutectic forms, still with falling temperature. Finally, all three constituents crystallise simultaneously at constant temperature, with the formation of a *ternary eutectic* (Fig. 32). An example of this in rocks is the intergrowth of quartz, orthoclase, and albite.

The above must be regarded as the discussion of an ideal case. In actual magmas there are several disturbing factors which have to be taken into account, such as super-cooling, pressure, presence of volatile constituents, and, above all, the abundance of mix-crystals (see next section). These factors may limit or inhibit the formation of eutectics.

MIXED CRYSTALS—In the discussion of the crystallisation of a binary magma it has been assumed that the constituents are of unchanging composition and independent of each other.   But the majority of pyrogenetic minerals are not of this character ; they consist of two or more components which are isomorphous and miscible in all proportions in the solid state, forming homogeneous crystals.   The plagioclases and the pyroxenes are especially good examples of this type of crystal ; and their formation from magmas follows totally different lines to those followed by constituents of constant composition.

FIG. 29.—GRAPHIC GRANITE.
Shows eutectic intergrowth between quartz (clear areas) and orthoclase (turbid areas) × 4.

The crystallisation of plagioclase felspar from a binary magma composed of the isomorphous albite and anorthite molecules presents an ideal example of mix-crystal formation.   The process may be represented, as before, by means of a temperature-composition diagram (Fig. 30).   In mix-crystals there is only one freezing-point curve (instead of two as in the case of non-mixing constituents), which is called the *liquidus* (AECB, Fig. 30).   Further, the melting-point curve, or *solidus* (AFDB), does not coincide with the freezing-point

5

curve. A mix-crystal does not melt at a definite temperature, but melting is spread over an interval of which the
lower limit is fixed by the solidus, and the upper limit by
the liquidus. At a temperature of 1400° C. the only liquid
that can exist has the composition C ($Ab_{65}An_{35}$), and the
only solid that can exist is one of composition D ($Ab_{25}An_{75}$).
In other words, only crystals of D composition are in equilibrium with a liquid C at that particular temperature.   It
follows that in a plagioclase magma the crystals formed at
any temperature are always much richer in the anorthite
molecule than the liquid with which they are in equilibrium.

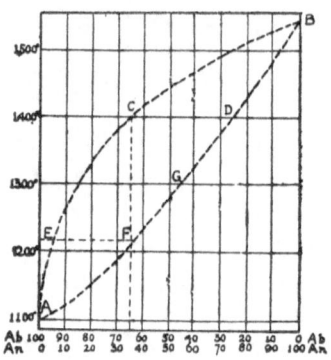

FIG. 30.—CRYSTALLISATION OF PLAGIOCLASE.
Temperature-composition diagram illustrating crystallisation of mixtures
of albite and anorthite (text, p. 65.)

Let a magmatic mixture of composition C now cool to 1400° C.
At this temperature crystals of composition D begin to form.
As the temperature continues to fall the liquid is enriched in
albite by the withdrawal of anorthite-rich crystals.   If the
cooling is very slow, so that equilibrium is constantly maintained between crystals and liquid, the crystals are continually made over or re-made to the composition appropriate
to the temperature, by reaction with the liquid.   At temperature 1300° C., for example, any new crystals formed
have the composition G ($Ab_{45}An_{55}$), and earlier ones have

THE FORMATION OF IGNEOUS ROCKS 67

been changed over to this composition. Crystallisation
ceases at the point E, when the crystals have the composition F, i.e. the same composition as the original liquid.
The final drop of liquid is of composition E (Ab$_{95}$An$_5$).

The continual complete readjustment of the composition
of the crystals in this process is only theoretically possible.
In actuality the adjustment is usually more or less incomplete. At a moderately rapid rate of cooling the crystals of
composition D are not changed over at the same rate as the
temperature falls. Hence zones are formed around a kernel,
each of composition appropriate to the temperature at which
it was formed, and to the liquid with which it was in equilibrium. Thus arises the well-known zonal structure of plagioclase crystals. Since less of the anorthite molecule is re-
turned to the liquid under these conditions, it follows that
the final liquid is richer in the albite molecule than in the
case of slow cooling. Zoning thus widely extends the range
of composition both of liquid and crystals. The kernels of
the crystals are rich in anorthite, and there is a progressive
enrichment in albite towards the exterior. The bulk composition, of course, will be that of the original liquid.

CRYSTALLISATION OF TERNARY MAGMAS—The crystallisation of a three-component magma can be represented by
means of a triangular diagram. Any mixture of three substances can be indicated by a point within an equilateral
triangle, such a point having the property that its distances
from the three sides are proportional to the amounts of the
three constituents. Thus in the triangle ABC (Fig. 31),
100 per cent. of A is represented by the point A, 100 per
cent. of B by the point B, and 100 per cent. C by the point C ;
i.e. unicomponent magmas are represented by the corners of
the triangle. Bi-component magmas are indicated by points
along the sides. Thus mixtures of A and B, B and C, C and
A, are found respectively along the sides AB, BC, and CA.
A mixture of three constituents is indicated by a point
within the triangle.

Let each side be divided into 100 parts, and let the amounts
of the constituents A, B, and C, be respectively a, b, and c
per cent. Then $a + b + c = 100$. To find the point representing this composition a distance BG $= c$ is marked off
from B, and a line GP is drawn from G parallel to AB. A

distance GP $= a$ is now marked off from G along the line GP. The point P is now the point required.[1]

The movement of P will now represent any possible change of composition of the magma. Movement towards A, B, and C means an increase in the relative amounts of these constituents respectively, and vice versa.

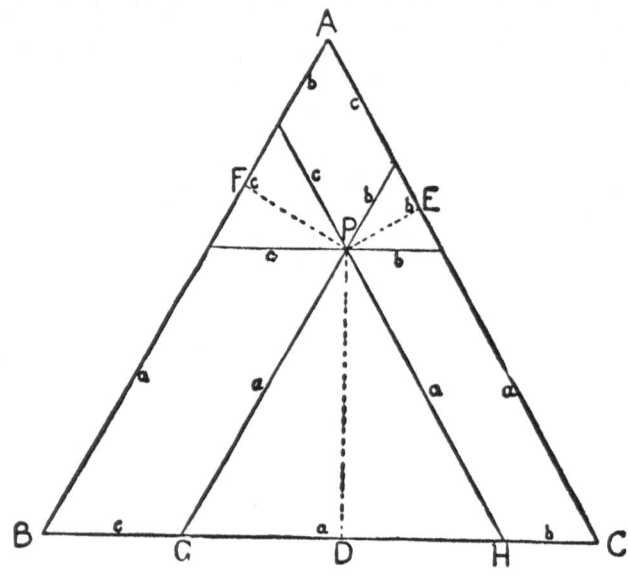

FIG. 31.—CONSTRUCTION OF TRIANGULAR CO-ORDINATES.
(Text, p. 67.)

To represent temperature ordinates perpendicular to the plane of the triangle must be erected, the ordinate at any point indicating the temperature at which the magma of that composition is saturated with the earliest crystallising constituent. Thus an ordinate at A indicates the temperature

[1] To prove that the distances of P from BC, AC, and AB respectively are proportional to a, b, and c, mark off CH $= b$ from C along CB. Then GH $= A$. The perpendicular PD then $= a\sqrt{3}/2$; similarly PE $= b\sqrt{3}/2$, and PF $= c\sqrt{3}/2$.

at which the pure substance A begins to crystallise. Similarly with B and C (Fig. 32). If this be done for all points on the diagram, a series of *temperature surfaces* will be obtained. The most convenient way of indicating these surfaces is to project their contours on to the plane diagram in exactly the same way that the undulating surface of the earth is projected on to a contoured map (Fig. 32).

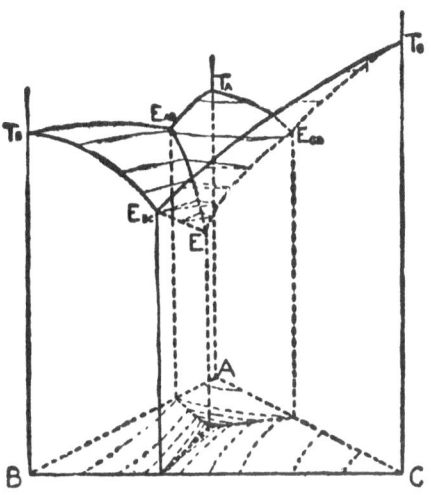

FIG. 32.—CRYSTALLISATION OF A TERNARY MAGMA.

T_a, T_b, and T_c represent the freezing-points of the pure substances A, B, and C respectively; E_{ab}, E_{bc}, and E_{ca} represent the binary eutectic points for the respective pairs indicated; E is the ternary eutectic for the three constituents. The freezing-surfaces are contoured, and the contour lines are projected on to the triangle at the base.

As an illustrative example of the crystallisation of a ternary magma the system diopside-albite-anorthite, worked out by the investigators of the Geophysical Laboratory at Washington, will be taken. Diopside is a pyroxene of composition $CaO,(MgFe)O, 2SiO_2$. Albite and anorthite are the end members of the isomorphous series of the plagioclases. Certain mixtures of these three components are closely similar in composition to the magmas which give rise to dolerites and basalts.

Every point in the diagram representing the crystallisation of this system (Fig. 33)[1] indicates a definite composition and a definite temperature. The triangle is divided by a curved line into two fields, each of which is crossed by temperature contours or isotherms. Diopside first crystal-

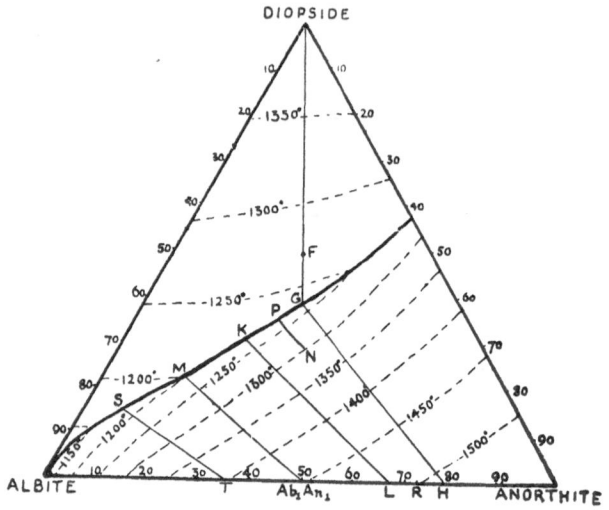

FIG. 33.—CRYSTALLISATION OF THE TERNARY SYSTEM DIOPSIDE, ALBITE, ANORTHITE.

(For explanation, see text, p. 70.)

lises from magmas of composition and temperature represented by points in the upper field, and plagioclases from magmas indicated by points in the lower field.

Let us follow the crystallisation of a magma of composition F, consisting of 50 per cent. diopside, 25 per cent. albite, and 25 per cent. anorthite. The point F representing this composition will stand half-way down the perpendicular

[1] N. L. Bowen, "The Later Stages of the Evolution of the Igneous Rocks," *Journ. Geol.*, 23, Supplement, 1915, pp. 33-9.

from the diopside corner, which bisects the base of the triangle. F also represents a temperature 1275°, which is that of the beginning of crystallisation; and as the point falls in the diopside field, diopside begins to crystallise at this temperature. By the withdrawal of diopside the liquid is enriched in albite and anorthite, and the indicator F will therefore move down the perpendicular until it reaches the boundary of the plagioclase field at G. At this point (temperature 1235°), the composition of the magma is 17 per cent. diopside crystals, 83 per cent. liquid (see table below). The plagioclase crystals separated at this point have approximately the composition Ab_1An_4, represented by the point H (i.e. much richer in An than is the liquid, see preceding section). The withdrawal of both diopside and anorthite-rich plagioclase from the liquid causes a change of composition indicated by the movement of the indicator point along the boundary curve to P, K, and M successively, the liquid becoming enriched in the albite component. If the cooling is slow enough to permit continuous and thorough adjustment of equilibrium, at P plagioclase of composition R separates out, and all earlier plagioclases will have been changed over to this composition. Similarly, at K the newly-formed and the earlier plagioclase crystals have the composition represented by L. The last drop of liquid is used up at M, and the plagioclase simultaneously arrives at the composition Ab_1An_1. The changes of temperature and composition at various stages are shown in the subjoined table.

Pt.	Temp.	Amount of Diopside.	Amount of Plagioclase.	Amount of Liquid.	Comp. of Plagioclase.
F	1275°	0	0	100	—
G	1235°	17	0	83	Ab_1An_4 (beginning of crystn.)
K	1220°	37	25	38	Ab_1An_2
M	1200°	50	50	0	Ab_1An_1

As diopside is a mineral of invariable composition in relation to the rest of the magma its composition remains unchanged; albite and anorthite, however, form mix-crystals which, as the table shows, vary continuously in composition during crystallisation. The remaining liquid also changes in composition, and the final drop is very much enriched in the albite component, as is shown by the position of the point M in the diagram.

Hence slow cooling and continuous adjustment of equilibrium between crystals and liquid results in the separation of crystals, diopside and plagioclase, in the proportions and of the compositions demanded by the known composition of the magma. With very rapid cooling the same final product is obtained. The liquid is greatly under-cooled, and crystallisation, when it takes place, occurs rapidly in the labile stage, giving 50 per cent. diopside, and 50 per cent. plagioclase of composition Ab_1An_1.

With the same magma an intermediate rate of cooling gives quite different results. Equilibrium between crystals and liquid is not perfectly adjusted, and the early crystals of anorthite-rich plagioclase are not transmuted into crystals of composition appropriate to the later magmatic stages. The earliest crystals are again Ab_1An_4; but as the liquid changes in composition with enrichment in albite, the next crop of plagioclase simply coats the early crystals with a zone of the appropriate composition at each stage. Hence, again, there is produced the well-known zonal structure. The composition of the remaining liquid is thereby rendered more extreme, since more of the anorthite molecule is permanently withdrawn from it. The result is just as if the plagioclase crystals formed at each stage were removed from the liquid, and the latter began to crystallise anew. Thus liquid of composition K, if crystallised without reference to previously formed plagioclases, would become completely crystalline at 1178° instead of 1200° as before, and the final liquid would have the composition S (Fig. 33), i.e. it would be much richer in albite than in the former case. Hence zoning causes a considerable widening of the range of composition, both of the plagioclase crystals and the liquid.

If the original magma is of composition N (diopside, 30 per cent. ; albite, 35 per cent. ; anorthite, 35 per cent.), crystallisation begins at 1302° ; and as the point N (Fig. 33) is in the plagioclase field, plagioclase of composition near Ab_1An_4 separates out. Hence the liquid is enriched in diopside and albite, and the indicator moves along the curve NP. At P diopside begins to crystallise, and the felspar has the composition R (Ab_1An_3). With further cooling crystallisation proceeds as in the above case; the last drop of liquid is used up at 1200°, and the felspar is made over entirely to Ab_1An_1. The crystallised mass now consists of 70 per cent. plagioclase (Ab_1An_1) and 30 per cent. diopside.

Owing to the presence of a mix-crystal series, there is in this process no appearance of simultaneous crystallisation of the constituents at constant composition and constant temperature, i.e. no eutectic is formed. Throughout the crystallisation there is a continuous change in temperature and composition.

CRYSTALLISATION OF OTHER TERNARY SYSTEMS—In the foregoing it is shown that complete adjustment of equilibrium between crystals and liquid consequent upon slow cooling, results in a rock consisting of minerals corresponding exactly in composition and amount with the known composition of the magma. If, however, equilibrium is disturbed, there is a great extension of the range of composition, and, in many systems, an increase in the number of minerals formed from the melt. One of the ways by which adjustment of equilibrium is hindered or inhibited is by zoning of early-formed crystals as explained above, with the formation of an armour which does not react with the liquid under the given conditions. Petrologists recognise two other conditions under which adjustment of equilibrium is prevented : removal of early-formed crystals from the liquid in which they were formed, by sinking under the influence of gravitation; and the straining-off, squeezing-out, or filtering, of the liquid from the crystals. The great majority of early-formed crystals are heavier than the magma, and, therefore, sink downwards at rates determined by their weight, shape, and the viscosity of the liquid. Bowen has shown that olivine crystals sink in laboratory melts ;[1] and numerous cases have been described of the accumulation of olivine and other heavy minerals towards the bases of igneous masses.[2]

---

[1] *Amer. Journ. Sci.*, 39, 1915, p. 175.
[2] J. V. Lewis (Palisade diabase), *Ann. Rept. State Geologist of N.J.*, 1907, p. 131 ; G. W. Tyrrell (Lugar sill), *Q.J.G.S.*, lxxii, pt. 2, 1917, pp. 84-131.

If earth movements involving compression occur during the crystallisation of a magma, it is conceivable that liquid of varying composition might be separated off at different stages from the crystals, and thus disturb equilibrium relations.

Another point brought out by the study of the diopside-plagioclase system, and by other systems involving mixed crystals, is that the earlier separations from any mix-crystal series are enriched in the component which has the highest M.P.; the residual liquid, and, therefore, the later crystals, are enriched in the less refractory constituent. The table below illustrates this point in the common mix-crystal

Mix-crystal Series.	Early Crystals Enriched in :	Residual Liquid and Later Crystals Enriched in :
Olivine, $2MgO, SiO_2$ — $2FeO, SiO_2$.	Mg-component.	Fe-component.
Orthorhombic pyroxenes, MgO, SiO_2 — FeO, SiO_2	Mg-component.	Fe-component.
Monoclinic pyroxenes, $CaO, MgO, 2SiO_2$ — $CaO, FeO, 2SiO_2$. (With molecules involving Al_2O_3, Fe_2O_3, TiO_2, Na_2O, etc.).	Mg-component.	Fe-component, and components with Al_2O_3, Fe_2O_3, TiO_2, Na_2O, etc. Also some enrichment in lime.
Plagioclases (Albite-Anorthite).	An-component.	Ab-component.

series of igneous rocks. This is supported by the observation that the ratio MgO : FeO in olivine or pyroxene rises with the proportion of olivine or pyroxene in the rock. The olivines and pyroxenes of gabbros are less rich in MgO relative to FeO than the olivines and pyroxenes of peridotites. This is because most peridotites are due to the segregation (and probably re-melting) of early-formed, and therefore Mg-rich, olivine crystals, from less basic magmas.[1]

[1] J. H. L. Vogt, " The Physical Chemistry of the Magmatic Differentiation of Igneous Rocks," *Videnskaps. Skr.* I *Math.-Nat. Kl.*, Kristiania, No. 15, 1924, pp. 9-52.

Similarly, the plagioclases of anorthosites (plagioclase-rich igneous rocks) are richer in the anorthite-component than the plagioclases of gabbros.

Other ternary systems which have been investigated in great detail are olivine-diopside-silica, and olivine-anorthite-silica. In bulk composition, mixtures of these components correspond to common magmatic types, such as gabbro, diorite, dolerite, and basalt. If cooling is slow, the early crystals are partly or wholly resorbed, resulting in a more or less uniform rock with a small range of composition. If, however, the early crystals, such as olivine or pyroxenes, sink out of the liquid; or if the liquid is strained off; or if the early pyroxenes or felspars become zoned; the range of composition of the liquid is greatly extended, with the result that minerals of widely different composition to the earlier ones can be formed. In the table given below, the early crystallisation and separation of the minerals in the left-hand column causes the concentration within the residual liquid of the substances in the right-hand column.

Early Crystals.	Residual Liquid.
Mg-rich olivine.	Iron-rich olivine and silica.
Mg-rich pyroxenes.	Iron- and lime-rich (diopsidic) pyroxenes and silica.
An-rich plagioclase.	Soda-rich plagioclase and silica.

The bodily separation and segregation of the early crystals in this case would, therefore, give rise to a gabbro-like rock. The residual liquid, on the other hand, would give rise by crystallisation to a granitic or granodioritic type rich in free silica. Hence arises the possibility of great variation in the rocks formed from the same initial magma, provided that the early crystals are separated off by some means from the residual liquid. This is the phenomenon of differentiation (Chap. VIII). Thus conditions (such as slow cooling and lack of disturbance) which favour continuous adjustment of equilibrium and resorption of early-formed crystals, produce uniform, average rocks of small range of composition. On the other hand, moderately rapid cooling or disturbances of equilibrium by various processes, hindering or preventing the absorption of early crystals, bring about a great diversity of igneous rock types from a common initial magma.

THE REACTION RELATION [1]—Once produced, a mineral formed as part of a eutectic system in magmas, is no longer concerned in the equilibrium. Its composition cannot be further changed; it is magmatically " dead." On the other hand, a mineral of a mix-crystal series is in continual reaction with the liquid from which it has crystallised, and its composition is continually being modified. The plagioclase series, as exemplified in Fig. 30, and the accompanying text, forms an ideal mix-crystal series. In their magmatic relations, Bowen proposes to call such series *continuous reaction series*.

Another type of reaction relation also occurs in magmas, in which an early-crystallised mineral reacts at a certain temperature with the liquid in such a way as to form a mineral of different composition. In the system $MgO-SiO_2$, for example, mixtures of appropriate composition produce olivine as the earliest mineral, but with falling temperature the olivine reacts with the liquid to form a pyroxene (clino-enstatite). The two minerals thus related by reaction are called a *reaction-pair*. A reaction relation of the same character may exist between three or more minerals which, when arranged in the proper order of succession, constitute a *discontinuous reaction series*. Minerals which are connected by this relation often exhibit *incongruent melting*, i.e. they have no definite M.P., but on heating break up into some other mineral plus liquid. Thus orthoclase at 1170° breaks up into leucite plus liquid, and clinoenstatite likewise breaks up into olivine and liquid.

Each mineral of a discontinuous reaction series may itself be a member of a continuous reaction series, and both kinds of reaction series may co-exist within the same magma. The reaction series believed to be present in common igneous

---

[1] N. L. Bowen, " The Reaction Relation in Petrogenesis," *Journ. Geol.*, 30, 1922, pp. 177-98.

rock types, in which ordinary basalt is assumed as the initial magma, may be represented as in the following scheme :—[1]

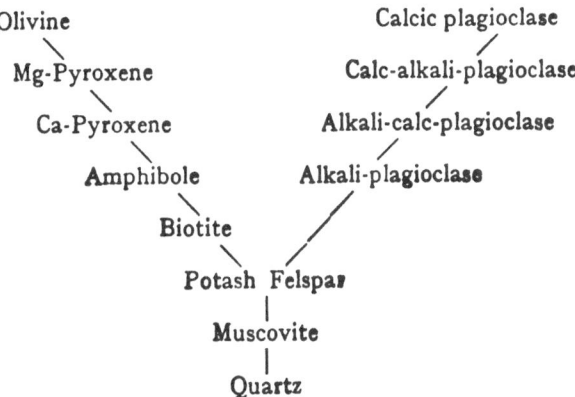

Olivine   Calcic plagioclase

Mg-Pyroxene   Calc-alkali-plagioclase

Ca-Pyroxene   Alkali-calc-plagioclase

Amphibole   Alkali-plagioclase

Biotite

Potash Felspar

Muscovite

Quartz

In this arrangement each mineral is supposed to react with the magmatic liquid so as to produce the mineral placed beneath it.   The left-hand branch is a discontinuous reaction series ;   the right-hand branch is the continuous reaction series of the plagioclases.   These branches converge, inter-lock, and finally merge into a single discontinuous series, of which quartz is the final product.

If reaction between a reaction pair is hindered or prevented, some constituent which would otherwise not be formed at all, is stored up in the liquid, and may appear as a mineral at a late magmatic stage.   Thus in the case of the system $MgO\text{-}SiO_2$ referred to above, the partial failure of reaction between olivine and liquid results in the enrichment of the liquid in silica, and the final crystallised product may be a mixture of olivine, pyroxene, and quartz.[2]   In this case the quartz is called a *released mineral*, and it is complementary to olivine, the mineral which normally should disappear by reaction.   Rocks containing released minerals are, in a sense, abnormal, because undersaturated minerals (such as olivine) may be present along with saturated or oversaturated

[1] N. L. Bowen, *op. cit. supra.*   [2] *Ibid.*, p. 182.

minerals (such as quartz).[1] Rocks in which released or *reactional* constituents (Lacroix) are prominent have been called *doliomorphic*.[2] An excellent example of a doliomorphic rock is a peculiar lamprophyre from Madagascar consisting essentially of biotite and quartz with a little hornblende. Its exceptional composition is explained as the result of the abundant early crystallisation of biotite, which failed to react with the liquid. Its formation used up only a part of the available silica, the rest of which was stored up in the liquid, and finally crystallised as quartz. Normally a magma of the composition represented by this rock would crystallise as a mixture of orthoclase and pyroxene, as shown by the following equation :—

$$2KAlSi_3O_8 + 2(Mg, Fe)SiO_3 = (Mg, Fe)_2K_2Al_2Si_3O_{12} + 5SiO_2$$
orthoclase + pyroxene   =   biotite   + quartz

The presence of olivine or analcite in rocks which, chemically, are saturated with, or have an excess of, silica, also provides examples of doliomorphism.

[1] See text, p. 104.
[2] A. Lacroix, "La notion de type doliomorphe en lithologie." *Comptes Rendus*, Paris, tom. 177, 1923, pp. 661-5.

CHAPTER V

TEXTURES AND MICROSTRUCTURES [1]

DEFINITION AND DESCRIPTION—*Texture* has already been defined (p. 33) as the intimate mutual relations of the mineral constituents and glassy matter in a rock made up of a uniform aggregate. It is best studied in thin section under the microscope. Microstructures, also studied in the same way, are due to the juxtaposition of two or more kinds of textural aggregates in a rock. Textures and structures are important, as these features are the indices of the geological processes which have been in operation; and their study provides valuable information as to the physical chemistry of the cooling and solidification of igneous rocks.

An accurate description of texture requires consideration of four points : (1) degree of crystallisation or *crystallinity ;* (2) absolute sizes of crystals, the *grain* or *granularity ;* (3) shapes of crystals ; and (4) mutual relations of crystals, or of crystals and glassy matter. The last two factors are sometimes grouped together as the *fabric.* The texture of an igneous rock may, therefore, be regarded as a function of its crystallinity, granularity, and fabric.

CRYSTALLINITY—Crystallinity is measured by the ratio subsisting between the crystallised and non-crystallised matter. A rock composed entirely of crystals is said to be *holocrystalline* (Fig. 34 A) ; when it consists wholly of glass, the term *holohyaline* is used (Fig. 34 c) ; and when the rock is composed partly of crystals and partly of glass, the terms

[1] A. Holmes, *Petrographic Methods and Calculations*, 1921, Chap. IX ; J. P. Iddings, *Igneous Rocks*, vol. i, 1909, Chap. VI ; A. Harker, *Natural History of Igneous Rocks*, 1909, Chap. XI. For the arrangement of this chapter I am especially indebted to Holmes.

mero-, *hypo-*, and *hemicrystalline* have been used (Fig. 34 B), of which mero is perhaps the best, as it has the widest connotation. The holocrystalline texture is characteristic of deep-seated or plutonic igneous rocks ; the merocrystalline

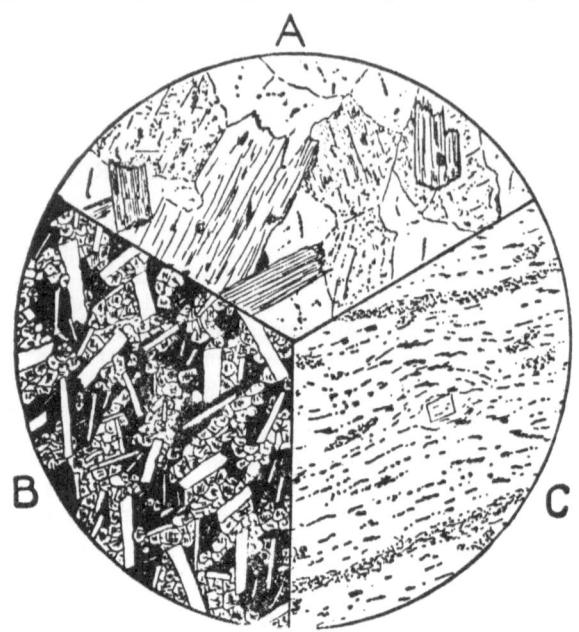

FIG. 34.—DEGREES OF CRYSTALLINITY.

A. Holocrystalline texture. Biotite-granite, Rubislaw, Aberdeen. Shows quartz, orthoclase, biotite, and magnetite.
B. Merocrystalline texture. Tholeiite, Brunton, Northumberland. Shows labradorite (laths), augite (granules), in a ground mass of dark glass.
C. Holohyaline texture, Obsidian, Lipari Is. Shows colourless glass with numerous microlites defining a flow structure, and one crystal of alkali-felspar
Magnification about 25 diameters.

of those which have consolidated on or near the surface. Holohyaline types are the least abundant, especially if crystallites and microlites (see below) are to be regarded as rendering the term holohyaline unsuitable. They occur most often as marginal facies of rock bodies, but may occur as lavas (obsidian), or as dykes and sills (pitchstone). A basalt dyke may exhibit each facies ; on the margins it may be glassy (tachylyte) ; in the centre it may be holocrystalline ; whilst in intermediate parts the texture may be mero-crystalline. The rate of cooling and the viscosity of the magma are the chief factors in determining the crystallinity, operating in the ways described in the last chapter (p. 59). Rapid cooling and high viscosity favour the formation of glass, whereas slow cooling and low viscosity promote the formation of crystals.

THE BEGINNINGS OF CRYSTALLISATION. CRYSTALLITES AND MICROLITES—In most natural glasses there are found

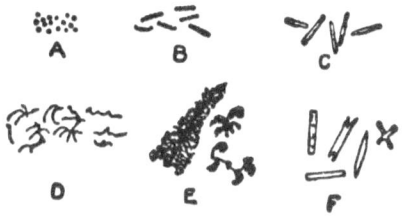

FIG. 35.—CRYSTALLITES AND MICROLITES.
A, globulites ; B, margarites ; C, longulites ; D, trichites ; E, scopu-lites ; F, microlites. Magnification about 200 diameters.

numbers of minute bodies of various shapes, which represent the beginnings of crystals (Fig. 35). *Crystallites* are embryo crystals, not yet organised to full crystalline status, and which therefore do not react to polarised light. *Microlites* are somewhat larger bodies which can be definitely recog-nised as minute crystals, whose properties are often deter-minable so that the mineral can be identified.

Microlites are usually rod- or needle-shaped, or exhibit the crystal outline appropriate to their mineralogical nature. Crystallites have more varied forms. *Globulites* are minute, spherical drops or pellets, often quite opaque, and then prob-ably consisting of iron oxide. When globulites are aligned and in contact, like a string of beads, the resulting form is called a *margarite*. *Longulites* are cylindrical rods with

6

rounded ends, which may be formed by the complete coal-
escence of a string of globulites. *Trichites* are filamentous
or hair-like forms which often spring from a common centre.
Scopulites are rods or needles carrying divergent plumes
or branches. These are well represented in some of the
Arran pitchstones, where they often pass into identifiable
crystals.

DEVITRIFICATION—Glasses are greatly super-cooled, excessively viscous solutions in which the molecules or atomic
groups are unarranged, and not in any definite order as in
crystals (p. 57). As this condition is unstable, glasses tend
to change in course of time, usually with the formation of
fibrous crystals. In other words, glasses are in a state of
deferred crystallisation. Hence very few geologically ancient
glasses are known; practically none are of pre-Carboniferous
age. And as glasses must have been formed in those remote
periods, it follows that they must subsequently have been
changed to crystalline material. The transformation of
glass to crystalline matter is called *devitrification*. Badly
made modern glass devitrifies in quite a short time. Thus a
200-year old glass recently dug up at Dumbarton was found
to be a mass of radiating crystals.

Devitrification is a very slow process at ordinary temperature and pressure, but heat, pressure, and circulating solutions, probably tend to speed up the process. Ancient glasses
have been subjected to heat and pressure by burial in the
crust, by compression in earth movements, and by igneous
intrusion. Heat is known to promote the spontaneous
crystallisation of glassy matter; and as the change from
glass to crystals involves reduction of volume, pressure must
also promote devitrification. That circulating solutions play
a part in the process is shown by the fact that, in ancient
glasses, devitrification often starts from joint fissures and
perlitic (p. 99) and other cracks, leaving cores of unaltered
glass in the centres of the affected blocks.

Devitrification in natural glasses usually produces a mass
of minute crystals of cryptocrystalline character (p. 83),
forming the felsitic texture (p. 86), and the rocks known as
felsites. It is not always possible to distinguish felsites
produced by the devitrification of glass from felsites formed
directly by the rapid cooling of granitic magmas. The

presence of perlitic cracks, however, is good evidence of the former glassy condition of the rock.

GRANULARITY—The absolute size of crystals in igneous rocks ranges from almost submicroscopic dimensions (e.g. microlites) to crystals measurable in yards, as in some pegmatites. This range may be more than 1 : 1,000,000 ; but in the average igneous rock the range of size is probably about 1 : 1000.

If the crystals are visible to the naked eye, or with the aid of a pocket lens, the rock is said to be *phanerocrystalline* or *phaneric*. If, on the other hand, the individual crystals cannot be distinguished in this way, the term *aphanitic* is used. Aphanitic rocks may be *microcrystalline*, i.e. the individual crystals are distinguishable with the aid of a microscope ; or they may be merocrystalline, crypto-crystalline, or glassy. The term *cryptocrystalline* is used when the individual crystals are too small to be separately distinguished, even under the microscope, and when their crystalline character is only recognisable by their confused aggregate reaction to polarised light. Phaneric rocks may be further distinguished as *coarse* when the average crystal diameter is greater than 5 mm. ; *medium-grained*, when it is between 5 mm. and 1 mm. ; and *fine-grained* when it is less than 1 mm.

The grain size of crystals depends on the rate of cooling and on the viscosity of the magma in just the same way as does the crystallinity, but in this case the molecular concentration of the crystallising substance in the magma is also an important factor (p. 60).

SHAPES OF CRYSTALS—The fabric or pattern of a rock depends on the shapes, and on the relative sizes and arrangement of the crystals   In this section the first of these factors is dealt with. Crystal forms are described with reference to the development of their faces as *euhedral*, when the crystal is completely bounded by faces (Fig. 36 B) ; and as *anhedral*, when crystal faces are absent (Fig. 36 A). The term *subhedral* is sometimes used for an intermediate stage of development (Fig. 36 c). Euhedral crystals are developed under circumstances, such as crystallisation in a thinly-liquid medium, of comparative freedom from interference by neighbouring crystals. Anhedral crystals, on the contrary, are of irregular shapes because their growth has been hindered by the juxta-position of other crystals, in consequence of which they have had to take the shapes of whatever open spaces were available between the already-crystallised minerals.

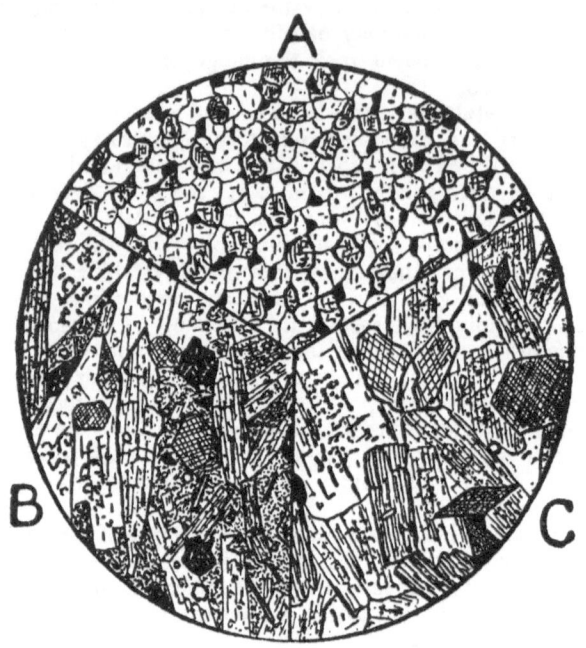

FIG. 36.—HOLOCRYSTALLINE TEXTURES.

A. Allotriomorphic texture.   Microgabbro (beerbachite), Odenwald.
   Shows anhedral crystals of labradorite, pyroxene, and iron-ore.
B. Panidiomorphic texture.   Lugarite, Lugar, Ayrshire.   Euhedral
   crystals of barkevikite, labradorite, ilmenite, and apatite, in a
   groundmass of turbid analcite.
C. Hypidiomorphic texture.   Syenite, Dresden.   Euhedral crystals of
   sphene and hornblende ; subhedral orthoclase and hornblende ;
   anhedral interstitial quartz.
   Magnification about 20 diameters.

Minerals such as olivine, hornblende, and biotite, are often found with more or less rounded outlines. The two last-named also frequently have dark corroded margins crowded with particles of iron oxide. These phenomena indicate that resorption, due to a change in conditions of equilibrium in the magma, has taken place.

Crystal shapes may also be described with reference to their relative dimensions in the three directions of space. *Equidimensional* crystals are those which are more or less equally developed in every direction, as, for example, polyhedral crystals such as augite, leucite, and garnet. Crystals better developed in two spatial directions than in the third, may be referred to as *tabular*. They form plates, tablets, flakes, scales, as, for example, in the micas and some felspars. Crystals better developed in one direction than in the other two are called *prismatic*. They form columns, prisms (thick and thin), rods, and needles. Hornblende and apatite are common examples of these forms. Finally, there are a number of crystal shapes, such as wisps, shreds, ragged patches, veins, and skeletons, which can only be described as *irregular*. Amongst the last-named may be mentioned the lattice-like skeletal crystals of ilmenite, and the peculiar embayed forms of quartz crystals in some quartz-felsites and porphyries.

MUTUAL RELATIONS OF CRYSTALS. EQUIGRANULAR TEXTURES—The fabric of a rock is influenced, not only by the shapes of the crystals, but by their relative sizes and mutual arrangements as amongst themselves, and with respect to any glassy matter which may be present. A little consideration shows that relative, and not absolute, size, affects the pattern; for the pattern would not be changed however much the absolute size might be magnified. Textures dependent on mutual relations may be classified as *equigranular*, *inequigranular*, *directive*, and *intergrown*, which are successively defined and described below.

Equigranular textures are those in which the constituent minerals are all of approximately the same size, so that the rock has an evenly-granular aspect, both in hand-specimens and in thin section. When most of the crystals are anhedral, the texture is further said to be *allotriomorphic* (Fig. 36 A), as in some aplites. In granites, syenites, and other plutonic rock types, most of the minerals are subhedral, and this texture is called *hypidiomorphic* or *granitic* (Fig. 36 C). The term *panidiomorphic* is used for textures in which the majority of the constituent crystals are euhedral, as in some lamprophyres (Fig. 36 B).

Equigranular textures also occur in microcrystalline rocks of hypabyssal and volcanic origin. In these fine-grained rocks the crystals are usually anhedral or subhedral, and the resulting texture is termed *microgranitic*.[1] In some highly felspathic rocks, such as orthophyres and plagiophyres, the felspars are mainly euhedral, giving rise to a fine-grained panidiomorphic texture which is called *orthophyric*. When the grain of a microgranular rock is so diminished that it becomes cryptocrystalline, a *felsitic* texture is produced ; but felsitic textures may also be developed by the reduction of all kinds of granitoid textures to the limit of microscopic visibility of the grains.

INEQUIGRANULAR TEXTURES—When the differences of size in the constituent minerals of an igneous rock become so pronounced that, megascopically or microscopically, they control the aspect of the rock, an *inequigranular* texture is produced. When the sizes vary gradually from the smallest to the largest, the fabric is further described as *seriate*. Usually there are two dominating sizes, with few or no crystals of intermediate size.

Two very important textures fall within this group, the *porphyritic*, and the *poikilitic*. In a porphyritic texture, the large crystals, or *phenocrysts*, are enveloped in a *groundmass* which may be microgranular, merocrystalline, or glassy (Fig. 37 A). In the last case the term *vitrophyric* is sometimes used to designate the texture ; and similarly the term *felsophyric* may be used when the groundmass is crypto-crystalline or felsitic. The texture may be described as *megaporphyritic*, and the large crystals as *megaphenocrysts*, when it is distinguishable with the naked eye. Similarly, the terms *microporphyritic* and *microphenocrysts* may be used when the texture can only be made out with the use of the microscope.

Poikilitic texture is the converse of porphyritic. The smaller crystals are enclosed in the larger ones without common orientation (Fig. 37 B). The inclusions must be numerous enough to produce a distinctive pattern, and therefore the minute inclusions, such as apatite needles and

---

[1] It may be suggested that the term *microgranular* would be more suitable. It would also conform with equi-, inequi-, intergranular, etc.

zircon crystals, which commonly occur within rock-forming minerals, are not regarded as conforming to the definition

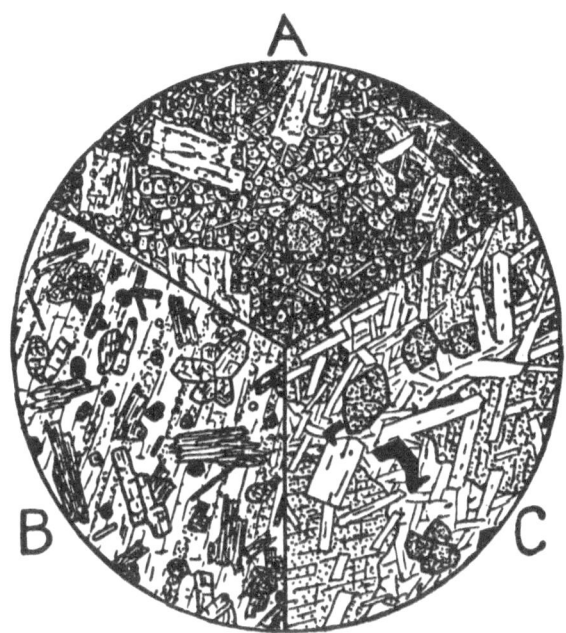

FIG. 37.—PORPHYRITIC, POIKILITIC, AND OPHITIC TEXTURES.

A. Porphyritic texture.   Olivine-basalt, Isafjord, Iceland.   Shows porphyritic labradorite, and olivine in a groundmass of augite, plagioclase, and iron-ores.   On the right hand is a glomeroporphyritic aggregate of labradorite, olivine, and ilmenite.

B. Poikilitic texture.   Shonkinite, Nordenskiöld Glacier, Spitsbergen. Shows euhedral augite, biotite, and apatite, enclosed in a large plate of orthoclase.

C. Ophitic texture.   Olivine-dolerite, Keflavik, Iceland.   Shows large plates of augite enclosing labradorite laths; other minerals are labradorite, olivine, and ilmenite.

Magnification about 20 diameters.

American authors have invented the terms *oikocryst* = (house-crystal) for the enclosing or host-crystal, and *chadacryst* (from the Gr. root, meaning to hold or contain) for the enclosed crystals.[1]

PORPHYRITIC TEXTURE [2]—This texture may arise in several different ways. The most important is that caused by a discontinuous change in the physico-chemical conditions attending crystallisation. The phenocrysts originate in depth, where the conditions of high pressure and slow cooling favour the formation of large crystals. The porphyritic texture is caused when the magma, with its enclosed crystals, is suddenly transferred to a higher level in the crust, or even extruded at the surface. With the diminution of pressure leading to loss of volatile constituents and increased viscosity, and the acceleration of the rate of cooling thus produced, large numbers of small crystals, or even glass and crystals, are produced, which form a groundmass to the earlier-crystallised phenocrysts. In this way arises the porphyritic texture of lavas and many minor intrusions.

Minerals whose constituent molecules are very abundant in the magma (i.e. high molecular concentration) must be relatively favoured as regards ability to grow large crystals. This condition may lead to seriate texture, especially if the magma becomes very viscous in later stages of crystallisation ; but the larger crystals of early formation may become enclosed in a groundmass of smaller ones. In this case there is no discontinuous change in conditions, and the phenocrysts may continue to grow, and thus be enabled to enclose some of the groundmass material along their margins. Further, the phenocrysts may not all have been formed at an early stage in the history of the magma. In many hypabyssal rocks the phenocrysts are not equally distributed over the whole igneous body, but are confined to the central or marginal parts. Similarly, they do not conform to any lines of flow which may be present, and occasionally take part in radiate or other groupings. These features indicate that the crystallisation of the phenocrysts took place along with the other constituents, and not before the emplacement of the intrusion.

Porphyritic texture may also arise from the fact that the phenocrystic mineral may have been relatively insoluble, and

[1] Iddings, *op. cit.*, p. 202.   [2] Holmes, *op. cit.*, p. 343.

began to crystallise well before the other constituents, thus being able to grow to a considerable size while the other minerals were still in solution. The growth of such minerals to large sizes is often aided by their rapid rates of crystallisation. Olivine in basalts is the outstanding example of this type of phenocryst. Large size would also follow from the probability that the phenocrysts develop mainly in the metastable stage, while the groundmass crystals tend to appear in the labile stage of crystallisation.

Porphyritic texture is confined largely to volcanic and hypabyssal rocks ; but some granites are also profusely porphyritic, as, for example, the Shap granite of Westmoreland, and the Loch Fyne granite of the south-west highlands of Scotland. The phenocrysts are usually alkali-felspars, the outer zones of which enclose small crystals of biotite, plagioclase, or magnetite. In some granites they are bordered by a zone of graphic intergrowth with quartz. The phenocrysts of granites are believed to represent an excess of the felspathic constituent over the eutectic proportion. Referring to page 63, it will be seen that crystallisation under these conditions takes place in two stages ; first the constituent which is in excess of the eutectic proportion crystallises, and then there is simultaneous crystallisation of the two main constituents. The porphyritic felspar then represents the first stage, and the granitic groundmass approximates to the second.

POIKILITIC TEXTURE—The conditions favouring poikilitic texture are complex and difficult to explain. One constituent must be present in relatively large amount as against a number of other constituents in relatively small amounts ; and this constituent must be the last to crystallise, yet under conditions permitting large growth. The whole crystallisation probably takes place within the metastable phase of cooling, and the comparatively small size of the earlier crystals can then be explained as due to their relatively small concentration. Their separation would cause the enrichment of the remaining liquid in volatile constituents, which would lower the viscosity, and tend to promote the growth of larger crystals. Poikilitic texture seems to be most common in rocks belonging to the syenites and monzonites, where orthoclase forms the host-mineral. It also frequently occurs in peridotites and picrites, in which horn-blende or biotite form the oikocrysts and olivines the chada-crysts.

The *ophitic* texture is a special case of poikilitic texture in which plates of augite enclose numerous thin laths of plagio-clase (Fig. 37 c). This is the characteristic texture of the rocks known as dolerite. With coarser grain, so that the felspars become of approximately the same order of size as the pyroxenes, their enclosure is only partial, and the texture is then referred to as *subophitic*. In some types the ophitic relation is confined to separated areas, between which there is a matrix free from augite and with an intersertal or intergranular texture. This appearance, which is really a structure as it involves the juxtaposition of two kinds of texture, has been called *ophimottling*.[1]

INTERSERTAL AND INTERGRANULAR TEXTURES—In most basalts the texture is determined by a framework or plexus of plagioclase laths or tablets so arranged that triangular or polygonal interspaces are left between the crystals. These interspaces may be entirely filled with granules of augite, olivine, and iron oxides, when the texture is desig-nated as *intergranular* (Fig. 38 c). Sometimes, however, glassy, cryptocrystalline, or fine-grained chloritic or serpentinous materials constitute important amounts of the infilling sub-stances, when the term *intersertal* is used.

DIRECTIVE TEXTURES—Textures which are produced by flow in magmas during their crystallisation are said to be *directive*. In so far as flow produces bands in which there is a juxtaposition of different textures, the appearances so caused have already been dealt with under the heading of structure (p. 39). But indications of flow may be present without destroying the uniformity of the rock. Thus crys-tallites, microlites, and crystals may be swung by flow into parallel lines or bands, which follow the stream-lines of the magma. If the magma was very viscous the stream-lines may be tortuous, and obstacles such as phenocrysts may produce much disturbance (Fig. 38 A).

Felspathic lavas, such as trachytes, phonolites, andesites,

---

[1] *Mull Memoir*, 1924, p. 138. The word *ophimottling* is rather an ugly hybrid ; the term *poikilophitic* seems to express the idea better.

etc., often have their felspar laths arranged in parallel position by flow. This texture is known as *trachytic* (Fig. 38 B). When the laths are interwoven with glass so as to produce a close

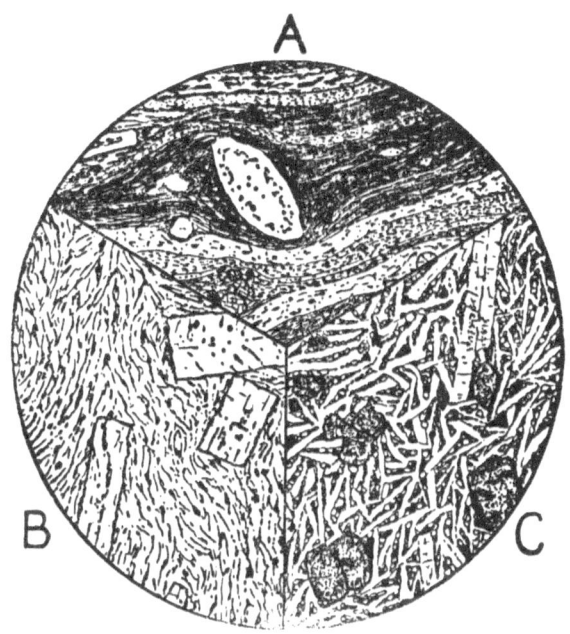

FIG. 38.—FLUIDAL, TRACHYTIC, AND INTERGRANULAR TEXTURES.

A. Fluidal structure.   Rhyolite, Iceland.   × 20.   Shows phenocrysts of alkali-felspar and augite in a flow-banded, glassy, and
cryptocrystalline groundmass.

B. Trachytic texture.   Bostonite, Great Cumbrae, Firth of Clyde.
× 20.   Shows phenocrysts of soda-orthoclase in fluidal groundmass of orthoclase laths, hæmatite, etc.

C. Intergranular texture.   Olivine-basalt, Kilpatrick Hills, Dumbartonshire.   × 60.   Shows microphenocrysts of labradorite and serpentinised olivine in a groundmass of labradorite, augite, and
iron-ores.

felted mass, the texture is said to be *hyalopilitic*. A more or less parallel arrangement of felspars in certain syenites is known as *trachytoid* texture. Phenocrysts of felspar and mica flakes in granites may occasionally be aligned in such a way as to indicate magmatic flow.

INTERGROWTH TEXTURES—Certain intergrowth textures have already been mentioned in connection with eutectic crystallisation (p. 63). The most common type is between quartz and felspars, the latter comprising orthoclase, microcline, perthite, and soda-plagioclase. Each mineral taking part in the intergrowth, notwithstanding apparent discontinuity, has the same optical orientation over large areas. In intergrowths involving quartz, this mineral is disposed in prismatic or wedge-shaped areas intersecting at angles of 60°, giving rise to the well-known *graphic* texture (Fig. 29). When this kind of intergrowth is reduced to microscopic dimensions the texture becomes *micrographic*, and the material is called *micropegmatite*. Intergrowths between quartz and felspars may not be so regular as in the graphic texture; irregular blebs, patches, and shreds of quartz are often seen in the felspars, and the texture is then referred to as *granophyric*. With reduction of grain size, a tendency to rough, radiate, or centric groupings of the quartz appears, which may pass in felsitic or rhyolitic types to radial aggregates of fibres, producing one kind of spherulitic structure (p. 98).

Intergrowths of orthoclase and albite (perthite and microperthite), and of orthorhombic and monoclinic pyroxenes, are known. In the first-named the albite often appears as irregular, elongated, vein-like patches enclosed in the orthoclase.

Many of these intergrowths are, no doubt, due to the simultaneous crystallisation of two mineral substances in eutectic proportions. Others, however, are not eutectic in origin, but are due to the phenomenon of *exsolution*.[1] In the case of microperthite, for example, it has been estimated that, at the temperature of crystallisation, orthoclase can hold in solution 28 per cent. of albite, but at ordinary temperature the solubility is only 10 to 15 per cent. Hence, on

---

[1] The term *exsolution* has been proposed by Alling to express the phenomenon of the separation of two crystal phases due to supersaturation. It is the opposite of " passing into solution." H. L. Alling, " Mineralography of the Felspars," *Journ. Geol.*, 29, 1921, p. 222.

cooling, orthoclase becomes supersaturated with albite, which separates out of solution in the solid state.[1]

REACTION STRUCTURES—In a previous section it was shown that reaction often takes place between early-formed minerals and the magma by which they are surrounded. The reacting mineral may entirely disappear; but occasionally the reaction is incomplete, and the corroded crystals are found surrounded by the products of the reaction. These are often grouped in such a way as to produce a texture which is

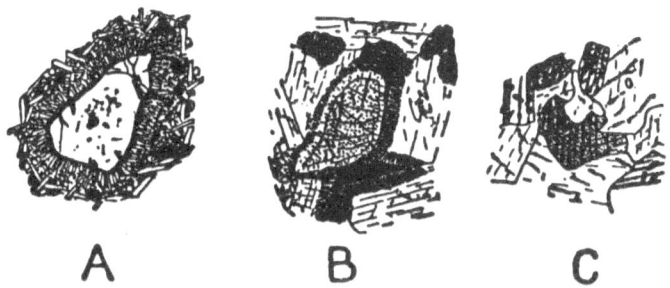

A   B   C

FIG. 39.—REACTION STRUCTURES.

A. Quartz grain enveloped in basalt lava, from the Lower Carboniferous, Mull of Kintyre. × 100. The grain is surrounded by a thick zone of prisms and granules of augite.

B. Corona structure. Olivine-norite, Risör, Norway. Shows olivine surrounded by a zone of colourless enstatite, and then (on the left) by an outer zone of green hornblende fibres The iron-ores are surrounded by a thick zone of brown hornblende fibres. × 20.

C. Myrmekite structure. Charnockite, Palni, Madras. × 20. Shows intergrowth of vermicular quartz with oligoclase.

different from that of the mass of the rock, and therefore the whole ranks as a structure. Reaction is frequently produced by the introduction of foreign minerals (xenocrysts) into a magma. Thus quartz accidentally introduced into a basaltic magma may be surrounded by a reaction zone of pyroxene granules (Fig. 39 A). Similar phenomena are believed to be due to interaction in the solid state between adjacent minerals, aided, in all probability, by the action of circulating fluids.

[1] A. Harker, *Natural History of Igneous Rocks*, 1909, p. 260.

The term *synantectic* has been applied by Sederholm to all mineral aggregates which have originated by these modes of reaction.[1] The zone of reaction products surrounding a mineral is called a *reaction rim*. Reaction rims may be further classified as *coronas* when produced by primary magmatic reaction, and as *kelyphitic borders* when secondary, paulopost,[2] magmatic action is concerned, or when the structure is due to reactions, such as those operating in metamorphism, which are independent of the original magma. Zoning of crystals and exsolution effects may also be regarded as reaction structures.

Corona structures often show a nucleus of an early-formed mineral such as olivine (in norites and gabbros), surrounded by one or more zones of reaction minerals, in this case, orthopyroxenes, or composite zones of two or more minerals.[3] These zones may be granular, or radiate-fibrous, in structure (Fig. 39 B). In the latter case, hornblende is often a constituent of the growth.

It is often impossible to distinguish between the primary corona structures and the secondary kelyphitic structures; and in such cases, the inclusive term reaction rim is preferable. Reaction rims are developed chiefly about olivine, hypersthene, and garnet, and the reaction minerals are usually pyroxenes and amphiboles, but occasionally a felspar or a spinel (magnetite or picotite) is formed.

Myrmekite is an intergrowth of quartz and plagioclase in delicate forms resembling micropegmatite, the quartz occurring as blebs, drops, and vermicular shapes within the felspar (Fig. 39 c). This growth takes place on the borders of orthoclase crystals, and appears to be due to the replacement of that mineral by plagioclase. This reaction involves the liberation of silica, which appears as quartz, and of potash, which goes to form the shreds of biotite often found in the neighbourhood of myrmekite. The growth is probably due, in most cases, to thermal metamorphism under uniform pressure.

XENOLITHIC STRUCTURES—Heterogeneity of texture is

---

[1] " On Synantectic Minerals," *Bull. Comm. Geol. de Finlande*, No. 48, 1916.

[2] From Lat. *paulo*, a little time; *post*, later.

[3] Harker, *op. cit.*, p 269.

produced by the inclusion of foreign rock fragments within an igneous rock (Fig. 52). This may arise in many ways, as, for example, in the emplacement of batholiths, sills, and dykes (p. 31), and the uprise of lava within a volcanic funnel. The inclusions, which may be of all sizes from chips of microscopic dimensions to blocks many yards in length, are called *xenoliths* (= stranger stones) or *enclaves* (Fr. = inclusions).[1] Xenoliths are called *cognate* when they represent fragments of rocks which are genetically related to the enclosing rock, and which, in most cases, have been formed at an early stage of crystallisation. *Accidental* xenoliths, on the other hand, are, as the name indicates, fragments of country rock, igneous, sedimentary, or metamorphic, as the case may be, which have been fortuitously included within the igneous rock as a consequence of its intrusion or extrusion. As examples of cognate xenoliths may be cited the olivine nodules which occur in some basalts, and the dark patches or segregations which are to be found in many granites. Olivine nodules are markedly angular, and appear to represent fragments torn from masses of olivine-rock. Bowen suggests that the formation of crystal-jams, or accumulations of early-crystallised minerals, as the result of clogging in the constricted parts of channels occupied by moving magma, is the cause of olivine-rock and other cognate inclusions in basaltic magma.[2]

There is a marked tendency for the phenocrysts in basalts—olivine, augite, labradorite—to segregate into clots which then form rocks identical with certain basic plutonic types. This structure was called *glomeroporphyritic* by Judd (Fig. 37 A).

The dark basic segregations of granites, in which the mafic minerals of granites are accumulated to a much greater degree than in the normal rock, probably also represent clots or patches of early-crystallised minerals included within the normal rock through some accidental circumstances arising during intrusion.

Accidental xenoliths, of course, may consist of any kind of

[1] A. Lacroix, *Les enclaves des roches volcaniques*, 1893.
[2] " Crystallisation-Differentiation in Magmas," *Journ. Geol.*, 27, 1919, p. 425.

rock; and owing to the frequent sharp contrast in composition between them and the magma by which they are enclosed, they often show notable contact-metamorphic effects.[1] Xenoliths are often partly or wholly absorbed or digested by the magma. Their outlines become rounded and indistinct (Fig. 52 c); finally, they are completely disintegrated, and their constituent minerals, altered and unaltered, dispersed throughout the surrounding magma. Their original positions are then marked only by slight variations in colour or texture, or by streakiness in the enclosing rock. Thus arise some types of *hybrid rocks*, so designated by Harker.[2]

ORBICULAR STRUCTURE—In some plutonic, and especially granitic, rocks, there occasionally occur ball-like segregations consisting of concentric shells of different mineral composition and texture, which may or may not be built up around a xenolithic kernel. The best known example is the famous orbicular diorite of Corsica. In this rock the centres consist of normal diorite, and are surrounded by alternating shells of radially-arranged bytownite felspar, and of felspar-poor mixtures of hornblende and pyroxene. In the Mullaghderg (Donegal) example, described by Professor G. A. J. Cole,[3] the kernels consist of schist xenoliths which have been partially or wholly replaced by granite. The products of this reaction diffused outwards, and formed a concentric ring consisting of radial andesine and magnetite, surrounding granite of ordinary texture, which immediately encloses the nucleus. Probably most examples of orbicular structure are to be explained as special cases of reaction with accidental xenoliths.[4]

SPHERULITIC STRUCTURE [5]—The essential feature of *spheru-*

---

[1] H. H. Thomas (Xenoliths, Mull), *Q.J.G.S.*, 78, pt. 3, 1922, pp. 229-60; N. L. Bowen, "The Behaviour of Inclusions in Igneous Magmas," *Journ. Geol.*, 30, 1922, pp. 513-70; H. H. Read (Xenoliths and Hybridism in Norite, etc., Aberdeenshire), *Q.J.G.S.*, 79, pt. 4, 1924, pp. 446-86.

[2] *Natural History of Igneous Rocks*, 1909, Chap. XIV. See also C. H. Clapp, "The Igneous Rocks of Essex Co., Mass," *Bull.* 704, *U.S.G.S.*, 1921, p. 115.

[3] *Sci. Proc. Roy Dublin Soc.*, xv, 1916, p. 141.

[4] Harker, *op. cit.*, p. 350; A. Holmes, *Petr. Methods and Calc.*, 1921, p. 360. See also J. J. Sederholm, "Orbicular Granites, etc.," *Bull. Comm. Géol. Finlande*, No. 83, 1928, 105 pp.

[5] Iddings, *Igneous Rocks*, vol. i, 1909, pp. 228-41; Harker, *op. cit.*, pp. 272-80.

litic structure is simultaneous crystallisation of fibres with radiate arrangement about a common centre (Fig. 40). The

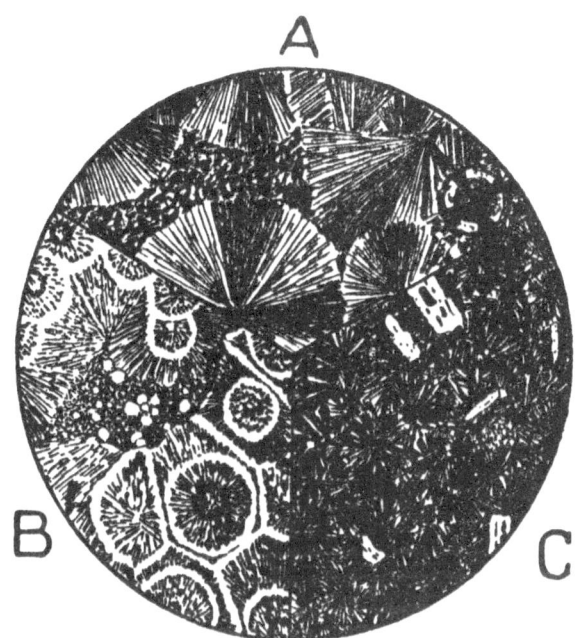

FIG. 40.—SPHERULITIC STRUCTURES.

A. Spherulitic pitchstone, Drapuhlidarfjäll, Iceland. Mostly radial felspathic microlites. One small concentric-radial spherulite on right. Patch of interstitial clay resulting from alteration of glass. Polarised light. × 20.
B. Spherulitic pitchstone, Corriegills shore, Arran. After Teall, *British Petrography*, 1888. Shows spherulites enclosed in polygonal cells. Interstitial patch of glass with small spherulites. × 20
C. Variolite, margin of tholeiite dyke, Dippin, Arran. Shows irregular sheaf-like spherulites of plagioclase, with a few microphenocrysts of labradorite and augite. × 20.

growth may be completely spherical, or it may consist of only a small sector of a sphere. Spherulites differ in their degree of perfection. The radiate arrangement may be extremely

7

regular, so that a perfect black extinction cross appears when the structure is viewed between crossed nicols under the microscope ; or it may form more or less irregular growths comparable to rough sheaves or bundles. Spherulites may or may not have a nucleus in the form of an early crystallised mineral. The growths may be completely separated ; they may be strung out as a line of coalescent spherules, thus starting apparently from a plane ; or they may be equidistant, and are then frequently enclosed in a polygonal cell which is due to the mutual interference of adjacent growths (Fig. 40 B). In addition to the radiate structure, spherulites often show concentric lines which appear to indicate interruptions at various stages of growth.

Spherulites may vary in size from microscopic dimensions to giant compound examples 10 feet in diameter (rhyolite of Silver Cliff, Colorado. Iddings, *op. cit.*). *Lithophysæ* (= stone bubbles) are large spherulites consisting of more or less complete concentric shells separated by empty spaces. They are ascribed to an alternation of crystallisation with the separation of a minutely-crystalline shell, and the emission of gas corresponding to the empty spaces.[1]

Spherulites mainly occur in the acid volcanic, or hypabyssal rocks, but they also appear, less abundantly, in basic lavas and intrusions. In the latter they are called *varioles*, and the rocks containing them, *variolites*. The name refers to the pitted, spotted, or minutely-nodular appearance of the surface of a weathered variolite.

In respect to mineral composition Harker (*op. cit.*) distinguishes two types of spherulites : (1) radiate growths of felspar fibres with interstitial quartz ; (2) radiate growths of felspar fibres only. Spherulites of the first kind probably represent eutectic intergrowths of quartz and felspar reduced to a microcrystalline or cryptocrystalline degree of fineness. On the exterior of a spherulite the radiate arrangement may pass into a distinct micrographic intergrowth (Harker, *op. cit.*, Fig. 90 A). In many cases, however, the felspar fibres dominate the growth, and are present in excess of eutectic proportions. The quartz may occur in fibrous forms or as granules packed in between the felspars. This type of

[1] F. E. Wright, *Bull. Geol. Soc. Amer.*, 26, 1915, p. 263.

spherulite is found in fine-grained granophyres, felsites, and pitchstones of intrusive origin.

Spherulites of the second type are abundant in rhyolites and other acid volcanic rocks, and in variolites. In the acid rocks they are composed of orthoclase fibres embedded in glass, or in cryptocrystalline or opaline silica, and with intercalated scales of tridymite (high temperature form of silica). In the variolites the felspars are plagioclase of sodic composition, which are often oligoclase.[1] Varioles are not so perfectly formed as spherulites, and show irregular extinction effects, due to oblique extinction and overlapping of the fibres or thin tablets (Fig. 40 c).

The fibrous habit suggests that spherulites developed rapidly in a highly supersaturated viscous solution. Fibres represent the crystalline form the production of which is most economical of energy under the prevailing conditions. Which of the two above-mentioned types of spherulite appears, depends on the composition of the magma. The localisation of spherulites probably depends on minute differences of physico-chemical character, especially the content of water vapour, and, therefore, viscosity, at certain centres, whereby crystallisation is favoured at these points. That spherulites often develop after the magma has come to rest, is shown by the fact that flow lines occasionally pass uninterruptedly through them ; or they occur as partial growths fringing subsequent cracks in the glassy matrix. The devitrification of a modern glass often results in spherulitic forms.

FRACTURE FORMS Some appearances in igneous rocks due to fracture may be classed as structures. In thin sections of volcanic glass curved concentric lines of fracture are often seen (Fig. 41 A). These are often so perfectly developed that, megascopically, the rock appears as an aggregate of little nodules consisting of a number of onion-like shells. The term *perlitic*, applied to these rocks, refers to these small spheres of glass, like pearls. Perlitic fracture is simply due to the contraction of the glassy mass on cooling. It can be imitated by allowing canada balsam to cool rapidly on a surface of roughened glass (Fig. 41 B).

[1] Tyrrell, " Variolites of Upper Loch Fyne and Skye," *Trans. Geol. Soc., Glasgow*, 14, 1913, pp. 291-302.

The only other fracture form it is necessary to notice here is the system of irregularly-radiating fissures which ramify through felspars and other minerals adjacent to

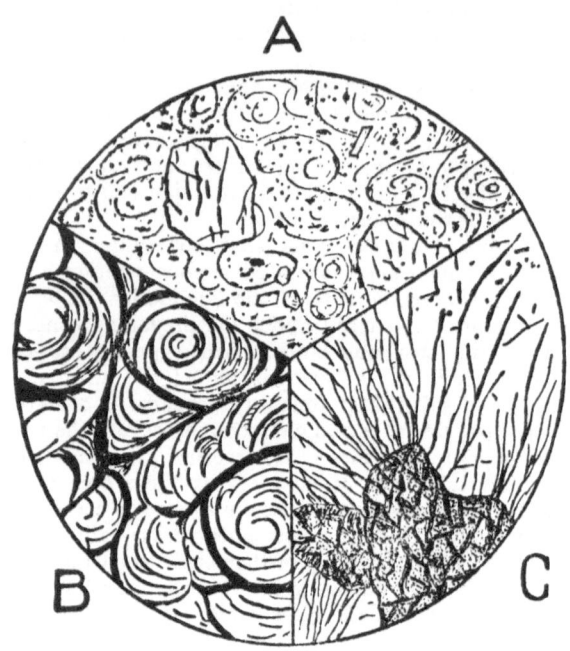

FIG. 41.—FRACTURE STRUCTURES.

A. Perlitic structure. Glassy rhyolite, Zarshuran, Persia.   × 20. Shows phenocrysts of alkali-felspar in colourless glassy groundmass with perlitic fractures.
B. Perlitic structure.   × 20. Artificially produced in canada balsam.
C. Expansion fissures, developed in plagioclase felspar around partially-serpentinised olivine. Troctolite, Volpersdorf, Silesia.   × 20.

altered olivine crystals in gabbros and norites. The alteration of olivine to serpentine involves a considerable increase in volume, and the stresses so produced are relieved by the fissuring of the surrounding minerals (Fig. 41 c).

CHAPTER VI

CLASSIFICATION OF IGNEOUS ROCKS

BASES OF CLASSIFICATION—There is no general agreement amongst petrologists as to the classification of igneous rocks. The lack of unanimity arises partly from the different bases of classification which have been proposed, and partly from misapprehensions as to the aims of classification. Igneous rocks show great variations, both in chemical and mineral composition; there are few, if any, definite natural boundaries or divisions within their range of composition. Hence the subdivisions have either to be comparatively broad, vague, and qualitative; or sharp, artificial, and quantitative, like the subdivision of the thermometric scale. Reconciliation of these divergent views may be brought about by the recognition that classifications on different lines, to serve different objects, may exist side by side. A distinction should be made between classification for a utilitarian end, such as research, description, comparison, teaching, field-work, and classification for ultimate truth (genetic classification), as methods and degree of elaboration will depend on the aim of the classification. We can have, for example, a *field* or *megascopic* classification, based on characters determinable in hand specimens; a *mineral* classification based on characters determinable by intensive laboratory work with the aid of the microscope and by other means; and a *chemical* classification based on composition as ascertained by chemical analysis. Then again, for *teaching* purposes, a simplified classification, based chiefly on mineral composition for reasons given later, must be available; and finally, after a sufficient amount of information has been accumulated, a true *genetic* classification, based on modes of origin, may arise. All these modes of classification may co-exist, self-supporting, and mutually explanatory, each useful for its own particular purpose.

There are four main bases of classification : *chemical, mineral, textural* (involving geological occurrence), and *associational* (involving geographical distribution and connection with geotectonic processes). Of these, only the first two are sus-ceptible to quantitative treatment.

THE FACTOR OF CHEMICAL COMPOSITION—Chemical classifi-cation, based on the chemical composition of the rocks, or of their hypothetical magmas, has lately been much in vogue. Since, however, it must be based on numerous chemical analyses, it is not adapted for the immediate determination, classification, and comparison of rocks. Furthermore, this basis makes the chemical composition the most fundamental character of igneous rocks. It may be so to the chemist ; but to the geologist the fact that rocks consist of minerals is the most important character, and classification based on mineral composition is, consequently, the most immediately useful for geological workers and students. Further, the determination of mineral composition is much more easily and quickly made than that of chemical composition.

An argument frequently used in favour of chemical classi-fication, as against mineral classification, is that the same magma may, under different conditions, give rise to totally different sets of minerals. Thus a magma which, under deep-seated conditions, and in the abundant presence of volatile constituents, may crystallise as an orthoclase-biotite rock (mica-syenite, or minette), will, under eruptive conditions and with loss of volatile constituents, give rise to leucite plus olivine (leucite-basalt). This fact, however, is more a reason for separating such different products in classification, as is done on the mineralogical basis, than for bringing them to-gether on the ground that they have similar chemical com-position. A chemical classification takes cognisance only of the magmatic composition ; the mineralogical classification takes account, not only of composition, but also of the cooling history of the rock. Nevertheless, chemical classification is essential and useful for many purposes, its greatest value being in the philosophical discussion of origins, and the indexing of rock analyses.[1]

---

[1] H. S. Washington, " Chemical Analyses of Igneous Rocks Published from 1884 to 1913," *Prof. Paper* 99, *U.S.G.S.*, 1917, pp. 1201.

In 1903 four American petrologists, Cross, Iddings, Pirsson, and Washington, advanced a classification which is in reality based on chemical composition.[1]   Analyses are first calculated into a set of standard minerals (the *norm*) under fixed rules which are based on the known laws of mineral formation in magmas.[2]   Certain important rock-forming minerals, the amphiboles, pyroxenes, and micas, are not utilised in the norm, because of their complex chemical composition. Their components are distributed between the norm minerals, which are based on magmatic molecules of fixed composition (p. 50).   The norm is divided into a *salic*[3] and a *femic* group, the most important constituents of which are the following :—

Salic Minerals.	Femic Minerals.
Quartz	Diopside
Orthoclase	Hypersthene
Albite	Olivine
Anorthite	Acmite (Aegirine)
Leucite	Magnetite
Nepheline	Ilmenite
Corundum	Hæmatite
Zircon	Apatite

The relative proportions of salic and femic minerals furnish the first line of subdivision into classes, and thereafter the method proceeds by taking factors from the norm two or three at a time ; but the full classification is too elaborate to be dealt with here.[4]   The conception of the norm is found to be of the greatest value in the comparison of igneous rocks.

Another quasi-chemical mode of classifying igneous rocks is by the use of the saturation principle (p. 49), as expounded by Shand[5] and Holmes.[6]   The minerals which are capable

[1] *Quantitative Classification of Igneous Rocks.*   Chicago, 1903.
[2] For the method of calculation, see Washington, *op. cit.*, pp. 1162-5 ; Holmes, *Petr. Methods and Calc.*, 1921, pp. 410-32.
[3] *Sal*ic, mnemonic for *sil*iceous, a*lum*inous.   *Fem*ic, mnemonic for *fer*ric or *fer*rous, *magnes*ian.
[4] For full description of the system see Washington, *op. cit.*, pp. 1151-61.
[5] "On Saturated and Unsaturated Igneous Rocks," *Geol. Mag.*, 1913, pp. 508-14 ; "A System of Petrography," *ibid.*, 1917, pp. 463-9.
[6] "A Mineralogical Classification of Igneous Rocks," *ibid.*, 1917, pp. 115-30.

of existing in igneous rocks in the presence of free silica are said to be *saturated ;* they are minerals of high silication. On the other hand, certain minerals of low silication, especially olivine and the felspathoid minerals (including analcite), are very rarely found in association with quartz, and may be termed *unsaturated*. These facts may form the basis of a classification as follows :—

 I. Oversaturated rocks, containing free silica of magmatic origin.

 II. Saturated rocks, containing only saturated minerals.

 III. Undersaturated rocks, containing unsaturated minerals.

  (*a*) Olivine-bearing rocks.

  (*b*) Felspathoid-bearing rocks.

  (*c*) Rocks with both olivine and felspathoids.

Thereafter the classifications of Holmes and Shand follow purely mineralogical lines.

Quartz is thus a *symptomatic* (Lacroix) or diagnostic mineral in over-saturated rocks, and olivine, felspathoids, analcite, etc., in undersaturated rocks. The doliomorphic rocks of Lacroix (p. 78), which contain released minerals due to incomplete reaction, or relics of early minerals which have escaped reaction, do not fit in well with this scheme of classification ; [1] but as they are comparatively rare, the point is of small practical importance. Hence, in regard to classification, little notice should be taken of small amounts of quartz, olivine, or felspathoids.

The *associational* factor in classification requires the discrimination of broad categories, suites, or kindreds of igneous rocks, which are *consanguineous*, as indicated by their constant and serial chemical and mineral characters, by their association in time and space (geographical distribution), and by their connection with special tectonic processes. This factor is dealt with in more detail in the next chapter.

THE FACTOR OF MINERAL COMPOSITION—Mineral composition is by far the most useful factor in classification for the research worker and student. In the first place, the actual units (minerals) of which rocks consist, are used ; in the

---

[1] A. Scott, " Saturation of Minerals," *Geol. Mag.*, 1914, pp. 319-24; " The Principle of Saturation," *ibid.*, 1915, pp. 160-4.

second, minerals can be quickly and easily identified and their relative amounts measured or estimated with a fair degree of accuracy. The most fundamental character of igneous rocks for the purposes of description and comparison is that they are aggregates of minerals, using the term mineral in its widest sense to include cryptocrystalline matter and glass (p. 7).

The minerals occurring in igneous rocks may be classed as *essential*, *accessory*, and *secondary*. The first two are products of magmatic crystallisation, and are therefore also *primary* or *original* minerals. The *secondary* minerals are those formed by weathering or metamorphism, or introduced by circulating solutions. *Essential* minerals are those which are necessary to the diagnosis of the rock type, and whose dwindling or disappearance would cause the relegation of the rock to another group. Minerals which are present in small amounts, and whose presence or absence is disregarded in defining the rock type, are called *accessory* minerals.

Another grouping of pyrogenetic minerals which has proved useful in classification is into felsic and mafic. *Felsic* is a mnemonic term derived from *fel*spar, *fel*spathoid, and *si*lica ; *mafic* is similarly derived from *ferro-magnesian*.

Felsic.	Mafic.
Quartz	Micas
Felspars	Pyroxenes
Felspathoids	Amphiboles
(including	Olivines
analcite)	Iron Oxides
	Apatite, etc.

The felsic group thus consists of minerals which are light in colour, of low specific gravity, and of comparatively late crystallisation ; the mafic group, on the other hand, comprises minerals which are dark in colour, heavy, and of comparatively early crystallisation. Rocks which are rich in felsic constituents are generally light in colour and of low density ; those with abundant mafic minerals are generally dark in colour and are heavy. Brögger's terms, *leucocratic* and *melanocratic*, referring primarily to light and dark aspect, may be qualitatively applied to rocks of richly felsic or richly mafic character respectively. Classification based on the relative proportions of felsic and mafic minerals is important, as differentiation, by whatever process it takes place, in general leads to the separation of the lighter, more siliceous, and more alkalic minerals, from the heavier ferro-magnesian minerals.

THE FACTOR OF GEOLOGICAL OCCURRENCE AND TEXTURE— The textural factor is used in most classifications after the mineralogical and chemical factors have been exhausted. It expresses the conditions under which cooling took place, that is, the geological occurrence. Igneous rocks have two chief modes of occurrence, *plutonic* and *volcanic*. Under plutonic conditions crystallisation takes place deepseatedly, with slow cooling under great pressure, and with retention of volatile constituents. Hence the textures of plutonic rocks are holocrystalline and coarse. Solidification under volcanic conditions takes place at low pressure, with loss of volatile constituents, and rapid cooling from a high temperature. The latter is partly original, and partly caused and maintained by oxidation reactions amongst the volatile gases, and between them and the atmosphere. The resulting textures are, therefore, mostly merocrystalline, or glassy, with numerous features, such as vesicularity and flow-structures, which are diagnostic of volcanic origin. The rocks are also frequently porphyritic.

Some petrologists admit only plutonic and volcanic modes of occurrence; but a third and intermediate mode, the *hypabyssal*, is recognised by many. The hypabyssal group includes the rocks of dykes, sills, and small laccoliths, etc., which occupy an intermediate position in the crust between the deepseated plutonic bodies, and the surficial lava flows. The textures of this group are naturally intermediate also, and range from holocrystalline to merocrystalline. As the occupation of higher levels in the crust means in most cases a sudden change in conditions, porphyritic texture is very common in hypabyssal rocks. This group, however, is not regarded as of equal importance with the plutonic and volcanic groups. It is often treated as a minor appendage of the plutonic group; but some hypabyssal rocks are certainly derived from volcanic magmas.

There is usually a slight but significant difference in chemical composition between plutonic types and the vol-

canic representatives of the same magmas, and this is connected with the very different cooling conditions operating
in each case. The plutonic rocks are comparatively rich in
minerals, such as hornblende and biotite, which require the
presence of volatile constituents and relatively low temperatures for their formation ; whereas volcanic rocks tend to be
richer in anhydrous high-temperature minerals such as the
pyroxenes. There is also greater opportunity for differentiation under plutonic conditions, so that extremely felspathic,
pyroxenic, hornblende-rich, and olivine-rich types are possible, which find no representatives amongst volcanic rocks.
Gabbros, for example, are richer in CaO and Al_2O_3, and
poorer in (Fe, Mg)O and Fe_2O_3, on the whole, than the corresponding basalts. Mineralogically, the result is a greater
richness in felspars, and poverty in ferro-magnesian minerals,
in gabbros than in basalts. This contrast is, no doubt, connected with more or less advanced differentiation in the
gabbroid magma, whereby some of the heavy mafic minerals
have sunk into the depths, leaving a magma relatively
enriched in felspars.

TABULAR CLASSIFICATION—The classification adopted in
this book is set out in the table below. Since felsic minerals
constitute 79 per cent. of the average igneous rock, the main
vertical subdivisions are based on the relations subsisting
between the three felsic mineral groups : quartz, felspars,
and felspathoids. As quartz does not occur along with felspathoids, this gives five divisions characterised respectively
by predominant quartz, quartz and felspars, felspars only,
felspars and felspathoids, and predominant felspathoids. To
these is added a sixth division for permafic or holomafic rocks
which are practically devoid of felsic minerals. Divisions I
and II belong to the oversaturated class of Shand and
Holmes (p. 104), III belongs to the saturated class, and IV,
V, and VI to the unsaturated class. This mode of arrangement also conforms fairly well with the silica percentage
mode of classification used by older writers, as is shown by
the silica percentages given in the table.

Horizontally classification is made into plutonic, hypabyssal,
and volcanic groups. In the plutonic group a further subdivision is made according to whether the rocks consist pre-
dominately of felsic minerals or of mafic minerals, or show

Classification table of igneous rocks.

	OVERSATURATED			SATURATED			UNDERSATURATED		
	I Quartz	II Quartz + Felspars		III Felspars			IV Felspars + Felspathoids	V Felspathoids	VI Mafic Minerals Predominant
		Predominant Orthoclase	Predominant Plagioclase	Predominant Alkali Felspars (Or, Ab)	Predominant Soda-Lime Plagioclase	Predominant Lime-Soda Plagioclase			
PLUTONIC — Felsic	Igneous Quartz Veins (Arizonite, Silexite)	Granite	Granodiorite (Tonalite)	Syenite	Diorite	Anorthosite	Nepheline-Syenite	×	
PLUTONIC — Mafelsic		×	×	×	×	Gabbro	Theralite and Teschenite ×	Ijolite	
PLUTONIC — Mafic				×	×	×	×	×	Peridotite Picrite
HYPABYSSAL		Granophyre Felsite	Pitchstone — Aplites — Porphyries — Lamprophyres			Dolerite / Tachylyte	Tinguaite		
VOLCANIC		Rhyolite	Dacite	Trachyte	Andesite	Basalt	Phonolite	Leucitophyre Nepheline-Basalt Leucite-Basalt	Olivine-rich Basalts
VOLCANIC (glassy)		Obsidian — Pitchstone	Pitchstone	Pitchstone		Tachylyte			Limburgite
Average Silica Percentage	90	72	66	59	57	48	54·5	43	41

108

approximate equality between the two groups; and in the table the names of the rock types have been placed in accordance with this subdivision. A cross indicates that rocks are known which occupy that place in the classification, although the names are not given. This mineralogical classification applies strictly only to the plutonic rocks. The volcanic rocks are placed in the table according to their crystallised minerals; but owing to the more or less abundant presence of glass and cryptocrystalline matter, the volcanic rocks may not be quite equivalent to the plutonic rocks appearing in the table above them. A further complication is introduced by the intrinsic and original difference in chemical composition between volcanic and plutonic magmas, referred to above. When the standard mineral composition of the average andesite, for example, is worked out, it is found to contain 15 per cent. excess silica, which must exist hidden or *occult* in the glassy or cryptocrystalline groundmass. If, therefore, an average andesite magma were to become holocrystalline, its mineral composition would be found to correspond more to quartz-diorite or granodiorite than to true diorite.

In the hypabyssal group certain textural types may have a wide range of mineral composition. Aplites and porphyries, for example, extend over practically the whole range. The composition of any particular type is indicated by prefixing the name of the plutonic rock to which it most closely corresponds, as granite-porphyry, syenite-aplite, etc.

NOMENCLATURE—While the nomenclature of igneous rocks has grown with the science, it has, unfortunately, developed in a casual and haphazard fashion, and has not been based on any definite scientific system. The terminations -*ite* and -*yte* are most used, along with roots which, according to Holmes,[1] have been drawn from no fewer than twenty-two different sources. These terminations, however, are also shared by minerals, alloys, explosives, wireless crystals, and a host of other natural and artificial substances. Some petrographical names are of ancient origin. *Syenite* (from Syene, Egypt), *basalt*, and *porphyry*, go back to Roman times. *Granite* was first used by Cæsalpinus in 1596. Other terms are adaptations of popular names or miners' designations in

[1] *Nomenclature of Petrology*, 1920, p. 2.

the countries where the rocks were first noted, as, for example, *gabbro*, from a Tuscan word ; and *minette*, from a term used by the Vosges miners. *Trachyte* is so named from the rough feel of the rock ; *clinkstone* (Hellenised to *phonolite*) refers to the musical ring of the rock under the hammer. Many of the newer names are based on the localities from which the rocks were first adequately described. With the exception of syenite, the first locality-name introduced was *andesite*, by L. von Buch (1835), for the typical lavas of the Andes. As further examples may be given *bostonite* (Boston, Mass.), *larvikite* (Larvik, South Norway), and *tonalite* (Tonale Alps). Another way of using a locality-name is to prefix the name of the locality to one of the major group names, indicating a special variety found and first described at that place. Examples are *Markle basalt* (Markle, East Lothian), *Ponza trachyte* (Ponza, Italy). Varieties of the principal rock types are often indicated by prefixing the name of some conspicuous mineral, as *biotite-granite*, *hypersthene-andesite*, *olivine-basalt*. For hypabyssal rocks textural terms, such as porphyry and aplite, are used, with the name of the corresponding plutonic or volcanic rock prefixed,[1] but many special names are also employed. Purely glassy rocks are indicated by special terms, such as *obsidian*, *pitchstone*, *tachylyte*, according to composition and appearance. For merocrystalline rocks with a comparatively large amount of glass, the prefixes *vitro-* and *hyalo-* may be used, vitro- indicating that the glass, while conspicuous, is present in smaller amount than the crystalline matter ; hyalo-, that the glass predominates.

GRANITE, GRANODIORITE, AND DIORITE—These three rock types form a natural association characterised by the presence of quartz, orthoclase, plagioclase, and various accessory minerals. Granite and granodiorite are rich in quartz (40 to 15 per cent.), but in quartz-diorite to diorite the amount of quartz gradually diminishes to zero (Fig. 42). From granite to diorite the proportional amount of orthoclase diminishes, while that of plagioclase (usually oligoclase) increases. The accessory minerals are most often biotite and hornblende, the biotite occurring mostly at the granite end, and the hornblende at the diorite end of the series.

---

[1] Tyrrell, *Geol. Mag.*, lviii, 1921, p. 499.

Pyroxenes (augite and hypersthene) are comparatively rare. Magnetite, pyrrhotine,[1] iron oxides, apatite, and zircon, are widely distributed, but in very small amount. The collective proportion of the mafic minerals increases from granite to

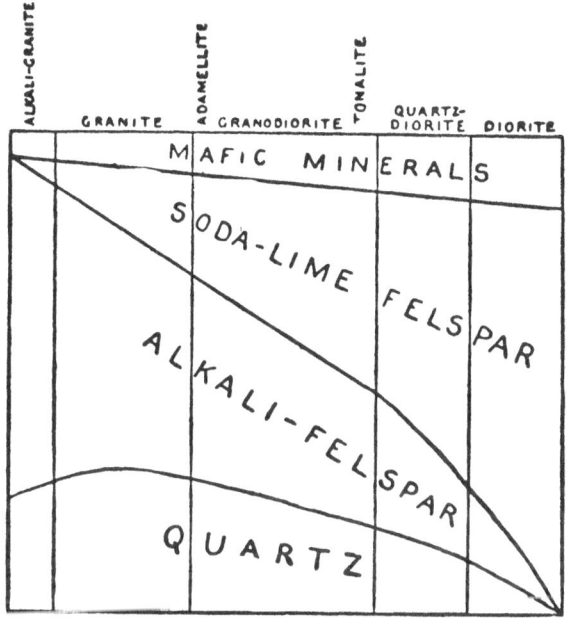

FIG. 42.—DIAGRAM ILLUSTRATING THE RELATIONSHIPS OF THE GRANITE-DIORITE SERIES.

The intercepts on any vertical line give the proportions of the constituent minerals for the rock represented.

diorite. The average chemical composition of rocks of this series is shown in Table IV. Corresponding to the variations in mineral composition, there is a decrease in SiO_2, Na_2O, and K_2O, and increase in Al_2O_3, iron oxides, MgO, and CaO, in passing from alkali-granites to diorites.

[1] Rastall and Willcockson, *Q.J.G.S.*, lxxi, pt. 4, 1917, p. 617.

TABLE IV

	1	2	3	4	5
SiO₂ . . .	73·2	72·0	66·0	59·5	56·8
Al₂O₃ . . .	12·5	13·1	15·3	16·5	16·7
Fe₂O₃ . . .	1·9	1·5	2·0	2·6	3·2
FeO . . .	1·3	1·8	2·9	4·1	4·4
MgO . . .	·2	·6	1·8	3·8	4·2
CaO . . .	·5	1·5	3·8	6·2	6·7
Na₂O . . .	4·6	3·5	3·5	3·0	3·4
K₂O . . .	4·9	4·8	2·8	1·9	2·1
H₂O . . .	·6	·7	1·0	1·4	1·4
TiO₂ . . .	·2	·3	·5	·6	·8
MnO . . .	·1	·1	·1	·1	·1
P₂O₅ . . .	—	·1	·3	·3	·2
Total . . .	100·0	100·0	100·0	100·0	100·0
Specific gravity .	2·64	2·64	2·73	2·80	2·85

1. Average alkali-granite, mean of 13 analyses.[1]
2. Average granite (hornblende-, biotite-, and augite-granites), mean of 19 analyses.[2]
3. Average granodiorite (tonalite, quartz-monzonite, and granodiorite), mean of 37 analyses.[3]
4. Quartz-diorite, mean of 20 analyses.[4]
5. Diorite mean of 70 analyses.[5]

In the normal granite type there is always a subordinate quantity of plagioclase, usually oligoclase. In *alkali-granites* this mineral disappears, and the mafic minerals are of alkalic varieties, such as aegirine and riebeckite, giving aegirine-granite and riebeckite-granite. Of the first, there is a good example in the rock of the isolated stack of Rockall in the North Atlantic. *Rockallite* consists of quartz, albite, and aegirine, in approximately equal amounts, and occurs as basic segregations in the aegirine-granite of Rockall.[6]

Normal granites are named according to the most prominent accessory mineral, as muscovite-, biotite-, hornblende-,

[1] Osann-Rosenbusch. *Gesteinslehre*, fourth edition, 1923, p. 113.
[2] *Ibid.*, p 116.
[3] Daly, *Igneous Rocks and Their Origin*, 1914, p. 25.
[4] *Ibid.*, p. 26.   [5] *Ibid.*, p. 26.
[6] Tyrrell, *Geol. Mag.*, lxi, 1924, pp. 19-25.

or augite-granite.   The Cornish and Scottish granites provide
numerous examples of the common biotite-, and hornblende-
bearing types.   Augite-granites are found among the Kainozoic intrusions of the Western Isles of Scotland, and hypersthene-granite (charnockite) is found in the Archæan of India
and West Africa.

With an increasing proportion of plagioclase relative to
orthoclase, granite passes to granodiorite.   There may also
be a slight diminution in the amount of quartz, and an aug-
mentation of the mafic constituents.   When the two felspars
occur in approximately equal quantity the term *adamellite* (Adamello Alps) is sometimes used. Similarly, *tonalite*
(Tonale Alps) is used for granodiorites in which the plagioclase is greatly in excess of the orthoclase.   Quartz-diorites
differ from the granodiorites in containing a much smaller
amount of quartz ;   otherwise they are very similar to tona-
lites.   With the practical disappearance of quartz and orthoclase, the series ends with diorite.

Granodiorites, quartz-diorites, and diorites, are abundant
among the Old Red Sandstone plutonic intrusives of Scotland.   Many of the so-called granite masses of Galloway, the
Grampians, and of the Aberdeenshire region, belong to this
series.

All these rock types from granite to diorite are characteristic plutonic associates of mountain-building activity.
Most of the great cordilleras of the present day, and the worn-
down mountain ranges of past geological times, show cores
of granodioritic types.   These rocks occur as huge batholiths, as well as in stocks and bosses.   Quartz-diorite and
diorite, however, tend to occur as marginal facies of granite
or granodiorite masses, and as small independent stocks and
bosses.

The use of the terms aplite and porphyry has already been
explained (p. 110).   An *aplite* is a fine-grained, equigranular,
allotriomorphic form of a plutonic rock, which occurs as
dykes, veins, or contact-facies and is usually slightly
differentiated from the parent magma in the " acid " or
felsic direction.   A *porphyry* is a hypabyssal form of a
plutonic magma, exhibiting one or more of the minerals
as phenocrysts in an aphanitic groundmass.   Nearly all
granite masses, and many granodiorites and diorites, have

8

associated aplites and porphyries in the form of satellitic dykes, and as marginal facies. The riebeckite-bearing rock of Ailsa Craig, in the Firth of Clyde, is an example of the aplitic form of an alkali-granite. An alternative method of naming aplites is to use the prefix *micro-* attached to the name of the plutonic rock to which they are related. Thus the Ailsa Craig rock may be called *riebeckite-microgranite.* A special but unnecessary name for this type is *paisanite* (Paisano, Texas). Similarly diorite-aplites or microdiorites are sometimes called *malchite.*[1]

When microgranite shows graphic intergrowths between the quartz and felspars it is called *granophyre.* The reduction of the grain-size to micro- or cryptocrystalline dimensions produces the type known as *felsite ;* and as quartz is frequently porphyritic in this type we have the rocks quartz-felsite or quartz-porphyry. Some types of pitchstone represent almost purely glassy forms of granite magma ; others correspond closely to granodiorite, or even diorite.

Syenite, Nepheline-syenite, and Related Alkaline Rocks—*Syenites* and *nepheline-syenites* are plutonic types which are characterised respectively by the predominance of alkali-felspar, and of alkali-felspar with nepheline. There may be a little accessory quartz and plagioclase in syenite ; and nepheline-syenites may carry a considerable proportion of albite, which ranks in this case as an alkali felspar. In syenite the mafic minerals may be hornblende, biotite, or augite ; hence there are hornblende-syenites, biotite-syenites, and augite-syenites. Pure orthoclase- or potash-syenites are extremely rare. In most cases the orthoclase is comparatively rich in the albite molecule, and there is quite an appreciable amount of plagioclase (oligoclase) and quartz present, as in the type syenite of Plauen, Saxony. Syenites which are very rich in alkali felspar and devoid of plagioclase are called *alkali-syenites.* Rocks of this character are the most abundant forms of syenite, and have been given many special names. With a little accessory quartz these rocks are called *nordmarkite* (Nordmark, Norway) ; with a slight deficiency of silica, so that a little nepheline makes its appearance, they are termed *pulaskite* (Pulaski, Arkansas). The beautiful

---

[1] From Malchen, the local name for Melibocus in the Odenwald.

" blue granite," so-called, which is so extensively used for shop fronts, belongs to the alkali-syenites and is called *larvikite* (Larvik, Norway). The shimmering blue felspar is anorthoclase ; the mafic minerals are titanaugite, biotite, olivine, and magnetite.

Syenites containing felspathoid minerals such as sodalite or analcite correspond chemically with the alkali-syenite group ; thus, for example, we have such types as the *sodalite-syenite* of Montana (U.S.A.), and the *analcite-syenite* of Mauchline, Ayrshire. Alkali-syenites, in which the proportion of mafic minerals is augmented so that the rocks belong to the mafelsic group, are called *shonkinite* (Shonkin Sag, Montana). Syenites and alkali-syenites in which plagioclase becomes approximately equal in amount to the alkali-felspars, are called *monzonite* (Monzoni, Tyrol). At the same time the plagioclase becomes more calcic, approaching andesine, or even labradorite, and the proportion of the mafic minerals is increased.

In the nepheline-syenite group the mafic minerals may be biotite, soda-pyroxenes (aegirine-augite and aegirine), soda-amphiboles (arfvedsonite, barkevikite, etc.), the garnet melanite, iron oxides, apatite, zircon, and sphene. In addition a great number of uncommon minerals occur in nepheline-syenites and their pegmatites, from which many rare elements have been obtained. Numerous special names have been given to varieties of nepheline-syenite, based on small differences, chiefly in the ferro-magnesian minerals. When albite is the only alkali-felspar present in the rock the name *canadite* has been used. Examples are found in Ontario, Sweden, and South Russia. The nepheline-bearing rock corresponding to monzonite is termed *nepheline-monzonite*, and has been described from Madagascar.

When plagioclase (usually labradorite) becomes the predominant felspar in nepheline-bearing plutonic rocks, we have the rock known as *theralite*. Rosenbusch's type theralite from Duppau, Bohemia, contains labradorite, nepheline, titanaugite, barkevikite, iron oxides, biotite, and olivine, and the rock is mafelsic in character.

The type corresponding to theralite, in which analcite takes the place of nepheline, is called *teschenite* (Teschen, Moravia), and is much more abundant than theralite. The greatest known development of teschenite is in the Midland Valley of Scotland, where it occurs as sills of Late Carboniferous and Permian age.[1] Differentiation from teschenite in the direction of concentration of the heavy minerals leads to the formation, first of *picrite*, which is a mafic type still retaining a little analcite and felspar, and finally *alkali-peridotite*.

Variation from nepheline-syenite in the direction of diminution and disappearance of all felspars, so that the rock consists of nepheline plus mafic minerals (chiefly soda-pyroxenes), leads to the type known as *ijolite* (Ijola, Finland).

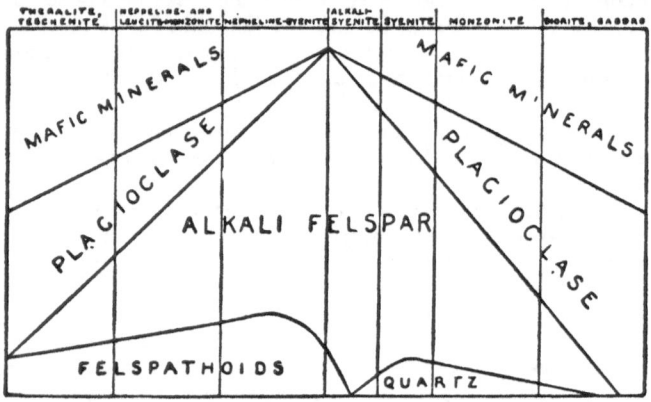

FIG. 43.—DIAGRAM REPRESENTING THE RELATIONSHIPS OF THE THERALITE—NEPHELINE-SYENITE—SYENITE—GABBRO SERIES

Ijolite is a mafelsic type ; the felsic rock in which nepheline predominates is called *urtite* (Lujaur Urt, Finland) ; and the corresponding mafic rock has been termed *melteigite* (Melteig, Fen, Norway).[2] The analcite-bearing rock corresponding to ijolite is *lugarite* (Lugar, Ayrshire), and to melteigite, *bekinkinite* (Bekinkina, Madagascar).

All the rocks described in this section occur as small stocks, volcanic plugs, sills, irregular intrusions, or large dykes.

---

[1] G. W. Tyrrell. "Classification and Age of the Analcite-bearing Igneous Rocks of Scotland," *Geol. Mag.*, lx, 1923, pp. 249-60.

[2] W. C. Brögger, *Eruptivgesteine des Kristianiagebietes*, iv, *Das Fen Gebiet.*, 1921, p. 18.

Their mineralogical relationships are exhibited in the diagram, Fig. 43, and their chemical characters in Table V.

TABLE V

	1	2	3	4	5	6
SiO_2 . . .	62·0	58·6	55·3	54 6	45·1	42·8
Al_2O_3 . . .	17·9	16·4	16·5	19·9	16·6	18·9
Fe_2O_3 . . .	2·2	3·6	3·0	3·4	4·8	3·9
FeO . . .	2·3	3·1	4·4	2·2	7·0	4·8
MgO . . .	1·0	3·1	4·2	·9	5·1	3·2
CaO . . .	2·5	4·5	7·2	2·5	9·3	10·5
Na_2O . . .	5·5	3·5	3·5	8·3	5·3	9·6
K_2O . . .	5·0	4·8	4·1	5·5	2·5	2·3
H_2O . . .	·8	1·1	·7	1·3	1·9	·8
TiO_2 . . .	·6	·9	·6	·9	1·7	1·6
MnO . . .	·1	·1	·1	·3	·1	·2
P_2O_5 . . .	·1	·3	·4	·2	·6	1·4
Total . . .	100·0	100·0	100·0	100·0	100·0	100·0
Specific gravity .	2·70	2·78	2·80	2·70	2·98	2·94

1. Average alkali-syenite, mean of 23 analyses. Daly, *Igneous Rocks and Their Origin*, 1914, p. 22.
2. Average syenite, mean of 11 analyses. *Ibid.*, p. 21.
3. Average monzonite, mean of 12 analyses. *Ibid.*, p. 23.
4. Average nepheline-syenite, mean of 43 analyses. *Ibid.*, p. 24.
5. Average theralite, mean of 4 analyses. Osann-Rosenbusch, *Gesteins-lehre*, 4th ed., 1923, p. 230.
6. Average ijolite, mean of 6 analyses. Daly, *op. cit.*, p. 33.

Porphyries and aplites of all these rocks do occur, but are comparatively rare. *Tinguaite* is a trachytoid dyke rock of the nepheline-syenite family, which is characterised by the abundance of needles of aegirine.

GABBRO, ANORTHOSITE, AND PERIDOTITE—The gabbros, along with anorthosite and peridotite, form a very important and abundant set of plutonic igneous rocks. Their mineral compositions and nomenclature are shown in the table on next page.

Biotite, hornblende, ilmenite, and magnetite, are accessories in gabbroid rocks, and are occasionally of sufficient local importance to form varieties. The texture is usually coarse, and hypidiomorphic. Anorthosites and anorthite-rocks are of felsic composition; pyroxenites and peridotites are mafic or holomafic; the remaining gabbroid rocks are usually mafelsic, the felspathic constituents being approximately equal in amount to the mafic constituents.

	Orth. Pyroxene.	Mono. Pyroxene.	Pyroxene and Olivine.	Olivine.	Felspars only.
Labradorite .	NORITE	GABBRO	OLIVINE-GABBRO	TROCTOLITE	ANORTHOSITE
Bytownite to Anorthite .	EUCRITE		ANORTHITE-PERIDOTITE	ALLIVALITE	ANORTHITE-ROCK
Labr. + quartz	QUARTZ-GABBRO				
+ nepheline and alkali-felspar	------ ESSEXITE ------				
Mafic minerals only . .	PYROXENITE		PERIDOTITE	DUNITE	

Norites, gabbros, olivine-gabbros, and troctolites, are abundant in Cornwall, the Ballantrae area (Ayrshire), Aberdeenshire, and the Western Isles of Scotland. The last-named region also provides the best-known examples of eucrite, anorthite-peridotite, allivalite, and anorthite-rock.

Quartz-gabbros appear to represent the plutonic form of a very abundant and widely-distributed magma, which has given rise to the extraordinarily numerous quartz-dolerite dykes and sills of Britain, the Eastern States of America, Guiana, South Africa, West Australia, etc.; quartz-gabbro or quartz-norite is also the staple rock of the great lopoliths of Sudbury (Ontario), and Duluth (Minnesota). In Britain the best example is that of Carrock Fell, in the Lake District. This rock consists of labradorite, diallage, and titaniferous iron ores, with quartz and alkali-felspar in micrographic intergrowth.

While quartz-gabbros are derived from normal gabbro magma slightly over-saturated with silica, *essexites* represent the same with a slight deficiency in silica, so that a small amount of nepheline replaces some of the felspar. Alkali-felspars may also be present, and the pyroxenes or hornblende may show characters indicating greater richness in soda than those of ordinary gabbros. In Britain typical essexites occur in connection with the Carboniferous igneous activity of Scotland. The type essexite of Essex Co., Mass., has, however, been shown to be partly a hybrid rock compounded of a mixture of gabbro and nepheline-syenite.[1]

In regard to origin *anorthosite* is a puzzling rock. It occurs as huge bodies of batholithic dimensions in Canada, the Adirondacks (New York), and in Scandinavia. Bowen believes that it is due to the segregation of plagioclase crystals from a gabbro magma, and that it never existed as an anorthosite magma.[2] Other authorities, however, especially Vogt, dispute this view.[3] The name was given by Sterry Hunt to the Canadian rock, and is based on the French term for plagioclase, *anorthose*. It is an unfortunate name, as it suggests that the rock is composed of anorthite or anorthoclase, whereas labradorite is the most usual felspar.

Pyroxenites and *peridotites* may be formed by the concentration of pyroxenes and olivine from alkalic rocks containing these minerals (p. 116); but usually they are associated with rocks of the gabbro family. Biotite and hornblende may occasionally be abundant, and give rise to special rock types. In many peridotites, chromite, or spinels such as hercynite and picotite, occur as accessory minerals. Numerous special names have been given to rocks of this group, based on small mineralogical differences. One of the best-known peridotites is that of the Lake of Lherz in the Pyrenees. It consists chiefly of olivine, with subordinate diopside, enstatite, and picotite, and is called *lherzolite*. The gabbro localities of Britain also provide many of the pyroxenites and peridotites.

The chemical composition of rocks dealt with in this section is illustrated by the selected analyses given in Table VI.

[1] C. H. Clapp, " The Igneous Rocks of Essex Co., Mass," *Bull.*, *U.S.G.S.*, No. 704, 1921, p. 124.
[2] N. L. Bowen, " The Origin of Anorthosite," *Journ. Geol.*, 25, 1917, pp. 209-43.
[3] J. H. L. Vogt, " The Physical Chemistry of the Magmatic Differentiation of Igneous Rocks," *Videnskapselsk. Skr. 1. Math.-Nat. Kl.* Kristiania, 1924, No 3, pp. 52-98.

TABLE VI

	1	2	3	4	5
SiO₂ . . .	53·8	48·2	48·6	50·4	41·1
Al₂O₃ . . .	16·8	17·9	18·0	28·3	4·8
Fe₂O₃ . . .	2·4	3·2	4·3	1·1	4·0
FeO . . .	7·0	6·0	5·6	1·1	7·1
MgO . . .	4·9	7·5	4·0	1·3	32·2
CaO . . .	7·0	11·0	8·9	12·5	4·4
Na₂O . . .	3·1	2·5	4·3	3·7	·5
K₂O . . .	1·6	·9	2·3	·7	1·0
H₂O . . .	1·7	1·4	1·3	·7	3·5
TiO₂ . . .	1·6	1·0	1·9	·1	1·2
MnO . . .	·2	·1	·2	—	·1
P₂O₅ . . .	·2	·3	·6	·1	·1
Total . . .	100·0	100·0	100·0	100·0	100·0
Specific gravity .	2·80 [1]	2·94	2·84	2·72	3·18

1. Quartz-gabbro, mean of 11 analyses of quartz-gabbro, quartz-norite, and quartz-orthoclase-gabbro, in Washington, *Chem. Anal. Ign. Rocks*, 1884-1913, *Prof. Paper* 99, *U.S.G.S.*, 1917.

2. Gabbro (including olivine-gabbro), mean of 41 analyses. Daly, *Igneous Rocks and Their Origin*, 1914, p 27.

3. Essexite, mean of 20 analyses. *Ibid.*, p. 30.

4. Anorthosite, mean of 12 analyses. *Ibid.*, p. 28.

5. Peridotite, mean of 31 analyses. *Ibid.*, p. 29.

DOLERITE AND LAMPROPHYRE—*Dolerite*, meaning a deceptive rock, was a term given by Haüy to certain dark, heavy, minutely-crystalline, igneous rocks of doubtful mineralogical character. It now denotes the hypabyssal forms of gabbroid, essexitic, theralitic, teschenitic, and basaltic magmas. Dolerite, in the restricted sense, means a rock composed of labradorite, augite, and iron oxides, the characteristic texture being ophitic. This composition corresponds to that of normal gabbro; the addition of olivine, or hypersthene, producing respectively *olivine-* or *hypersthene-dolerite*, gives rocks corresponding to olivine-gabbro and norite. Accessory quartz, often intergrown with alkali-felspars, gives the type *quartz-dolerite*, corresponding to quartz-gabbro. This is an exceedingly abundant rock

[1] Carrock Fell.

all the world over,[1] and forms the staple rock of the great east-west dykes and accompanying sills of the Midland Valley of Scotland, and of the North of England (the Great Whin Sill). Doleritic rocks representing the hypabyssal forms of essexite, theralite, and teschenite, are also known, and may be termed respectively *essexite-*, *theralite-*, and *teschenite-dolerite*. The chemical composition of dolerite very closely corresponds to that of its parent plutonic type.

Dolerite, in which the plagioclase has been partially or wholly altered with the formation of carbonates, zoisite, epidote, or albite, the pyroxene changed to chlorite or hornblende, the olivine to serpentine, and the ilmenite to leucoxene, is called *diabase*. This term tends to become obsolete in British petrographical literature, but it should be retained, as it denotes a very common type of rock which is especially abundant in connection with metalliferous ore deposits, and in regions of low-grade metamorphism.

The term *lamprophyre* was first applied by von Gumbel to certain melanocratic dyke rocks of the Fichtelgebirge (Saxony). The group covers a wide range in chemical and mineral composition. The rocks are mostly dofelsic or mafelsic, and are characterised by an abundance of euhedral crystals, producing the panidiomorphic texture. They occur as dykes and small sills. The constituents are felspars, including both orthoclase and plagioclase (oligoclase or andesine), and ferro-magnesian minerals which may be pyroxenes or amphiboles, but are most frequently biotite. Olivine may occur in more basic types, and is generally serpentinised. The composition and nomenclature of the chief types are shown in the table on next page.

Minette, vogesite, kersantite, and spessartite, occur as basic differentiates of granitic or granodioritic magmas, and are complementary to aplites and pegmatites. They are especially abundant in the mountain fragments of Late Carboniferous times (Hercynian), which are found in Cornwall and Devonshire, Brittany, South Germany, and Saxony, whence the names are derived. Soda-minette, camptonite, alnöite, and monchiquite, are genetically connected with alkali-rich plutonic types, especially nepheline-syenite.

[1] G. W. Tyrrell, *Geol. Mag.*, 1909, pp. 299-309; 359-66.

Ferro-magnesian Minerals.	Orthoclase Predominant.	Plagioclase Predominant.	Felspar-free.
Biotite	MINETTE (Miners' term in Alsace)	KERSANTITE (Kersanton, Brittany)	(With melilite) ALNÖITE (Alnö, Sweden)
Augite and/or Hornblende	VOGESITE (Vosges Mts.)	SPESSARTITE, ODINITE (porphyritic var.) (Spessart Mts., Germany)	
Alkalic pyroxenes and/or Amphiboles	SODA-MINETTE (Also vogesite in part)	CAMPTONITE (Campton Falls, New Hampshire)	(With analcite) MONCHIQUITE (Serra Monchique, Portugal)

RHYOLITE AND DACITE—*Rhyolite* is the volcanic equivalent of granite. It exhibits the same minerals as granite, embedded in an abundant glassy or cryptocrystalline groundmass which is very frequently flow-banded. The crystalline components are often feebly developed or absent. The extreme glassy modification of rhyolite is the pure black glass known as *obsidian*. Many pitchstones, too, are of rhyolitic composition, and are to be distinguished by their pitchlike lustre. The name rhyolite, given by von Richthofen in 1861, has reference to the frequent occurrence of flow structures in these rocks. The term *liparite* (Lipari Islands) is a synonym for rhyolite which was proposed by J. Roth, also in 1861, and is still much used by Continental petrographers. Many ancient rhyolites, obsidians, and pitchstones, now have the texture of *felsite*, owing to the tendency of glass to devitrify in course of time. Their original nature is often betrayed by relics of spherulitic, perlitic, and flow structures.

Rhyolites which carry a notable proportion of soda-lime plagioclase, and are thus equivalent to the adamellite variety of the granodiorites, are called *dellenite* (Dellen, Sweden).[1] Owing to the frequent difficulty of discrimination, dellenites

[1] G. W. Tyrrell, " Contribution to the Petrography of Benguella, ' *Trans. Roy. Soc. Edinburgh*, 51, pt. 3, 1916, pp. 537-59.

have not often been described as such; but it is probable that they have been regarded as rhyolite or dacite, and are much more abundant than is supposed. Those rhyolitic types in which the plagioclase, visible or occult in the groundmass, considerably exceeds the orthoclase, and which, therefore, correspond to tonalite and quartz-diorite, are called *dacite* (Dacia = Hungary). Dacites approximate to andesites in composition and texture, but are distinguished therefrom by the abundance of quartz, either as phenocrysts or in the groundmass.

Some types of rhyolite are the volcanic equivalents of soda-granites, and carry phenocrysts of soda-orthoclase or anorthoclase, and crystals of soda-pyroxenes and soda-amphiboles, indicating their alkalic characters. *Comendite* (Comende, Sardinia) is an alkali-rhyolite with aegirine, arfvedsonite, and riebeckite. *Pantellerite* (Pantelleria, Mediterranean) is distinguished by the presence of anorthoclase, with aegirine-augite, and the triclinic alkali-amphibole cossyrite.

In Scotland rhyolite is occasionally found associated with andesite and dacite in the Old Red Sandstone lava fields, and very sparingly in the Kainozoic igneous suite of the Western Isles. Explosive episodes in the latter are due to the extrusion and intrusion of rhyolitic rocks. In Wales the Older Palæozoic igneous suites show many rhyolites, which are now thoroughly devitrified and often partially silicified. Obsidian is characteristic of several recent volcanic outbreaks, such as those of Iceland, the Yellowstone Park, and New Zealand. The alkali-rhyolites have been found in several Mediterranean localities as is denoted by their names. They also occur in Kenya Colony, Somaliland, Socotra, and many other African localities. In Britain soda-rhyolites have recently been described from Skomer Island, Pembrokeshire. Dacite is generally found in the great andesite fields (p. 127).

The average chemical composition of members of this group is displayed in Table VII, p. 124.

TRACHYTE AND PHONOLITE—The term *trachyte* was formerly applied to all lavas which felt very rough to the touch, but is now restricted to the volcanic equivalents of the syenites, i.e. to lavas which are rich in orthoclase or other alkali felspars. The characteristic felspar may occur as

TABLE VII

	1	2	3
SiO$_2$ . . .	72·1	72·6	66·9
Al$_2$O$_3$ . . .	10·2	13·9	16·6
Fe$_2$O$_3$ . . .	3·4	1·4	2·5
FeO . . .	2·4	·8	1·4
MgO . . .	·3	·4	1·2
CaO . . .	·4	1·3	3·3
Na$_2$O . . .	5·0	3·6	4·1
K$_2$O . . .	4·5	4·0	2·5
H$_2$O . . .	1·1	1·5	1·1
TiO$_2$ . . .	·4	·3	·3
MnO . . .	·1	·1	—
P$_2$O$_5$ . . .	·1	·1	·1

Summation, 100.

1. Alkali-rhyolite, mean of 13 analyses of comendite and pantellerite. Osann-Rosenbusch, *Gesteinslehre*, 1923, p. 366.

2. Rhyolite, mean of 64 analyses of rhyolites and liparites. Daly, *Igneous Rocks and Their Origin*, 1914, p. 19.

3. Dacite, mean of 30 analyses. Daly, *ibid.*, p. 25.

phenocrysts (often the sanidine variety of orthoclase), in a groundmass which consists largely of flow-orientated microlites of alkali felspar (trachytic texture). Trachytes are named according to the nature of the predominating coloured mineral as biotite-trachyte, augite-trachyte, hornblende-trachyte, etc. Volcanic rocks corresponding to the soda-syenites are common, and are known as *soda-trachytes*. They include varieties with aegirine or riebeckite as the prominent mafic minerals ; or with anorthoclase, titanaugite, occasionally olivine, and a little nepheline, when the name *kenyte* (Kenya) is applied. Other varieties carry a small quantity of analcite or nepheline, and represent transitions to the phonolites (*phonolitic trachyte*).

The volcanic rock which is the equivalent of monzonite, carrying alkali-felspar and soda-lime felspar in approximately equal amounts, is called *trachyandesite*. Augite, biotite, or hornblende are the most common mafic minerals, and the rock usually has a trachytic texture. *Latite* (Latium) is a practically synonymous term. Obsidians and pitchstones of trachytic and trachyandesitic composition are known, but are not so abundant as those of rhyolitic composition.

Trachytes and trachyandesites are represented in many oceanic islands (Ascension, Samoa, Jan Mayen, Hawaii, etc.), and in mildly alkalic petrographic provinces such as the Carboniferous of Scotland, and the Kainozoic igneous fields of Germany and France (Auvergne), as differentiates from basaltic magma.

Phonolite is the Hellenised form of the term clinkstone (due to Werner), which was applied to certain German lavas whose platy fragments gave a musical ring under the hammer. In modern usage, however, the term phonolite implies a lava characterised by essential nepheline or leucite, thus corresponding in composition to nepheline-syenite or leucite-syenite. The nepheline-bearing types are by far the most abundant, and carry alkali-felspar in addition to the nepheline, along with soda-rich pyroxenes and amphiboles. Some phonolites are rich in other felspathoids such as nosean and haüyn, and are therefore called *nosean-*, and *haüyn-phonolite*. When a leucite-phonolite becomes very rich in leucite it is called *leucitophyre*. The phonolitic lava which carries essential plagioclase as well as nepheline and alkali-felspar, and thus corresponds in composition to nepheline-monzonite, has been called *vicoite* (Vico, Italy).

Phonolites occur sporadically in the Carboniferous of Scotland (Traprain Law; Eildon Hills; Fintry). A very fine example of nosean-phonolite comes from the isolated Wolf Rock off the Cornish coast. The typical localities for phonolite are in Saxony, Bohemia, and South Germany. Leucite-phonolites and leucitophyres are abundant in the Eifel and in several Italian provinces.

The average chemical composition of members of this group is shown in Table VIII, p. 126.

ANDESITE AND BASALT—Andesite and basalt are very important groups as the basalts are the most abundant, and andesites the next most abundant, types of lavas (p. 54). Their abundance renders their exact discrimination a matter of some importance.[1] Calculation of the standard mineral

---

[1] G. W. Tyrrell, "Some Points in Petrographic Nomenclature," *Geol. Mag.*, lviii, 1921, pp. 496-9.

TABLE VIII

			1	2	3	4
SiO$_2$	.	.	.	60·7	62·0	57·5	56·2
Al$_2$O$_3$	.	.	.	17·7	17·5	20·6	17·4
Fe$_2$O$_3$	.	.	.	2·7	2·9	2·4	3·7
FeO	.	.	.	2·6	1·6	1·0	3·4
MgO	.	.	.	1·1	·6	·3	2·5
CaO	.	.	.	3·1	1·1	1·5	5·2
Na$_2$O	.	.	.	4·4	6·6	8·8	5·0
K$_2$O	.	.	.	5·7	5·9	5·3	3·6
H$_2$O	.	.	.	1·3	1·2	2·0	1·3
TiO$_2$	.	.	.	·4	·4	·4	1·0
MnO	.	.	.	·1	·1	·1	·2
P$_2$O$_5$	.	.	.	·2	·1	·1	·5

Summation, 100.

1. Trachyte, mean of 48 analyses. Daly, *Igneous Rocks and Their Origin*, 1914, p. 21.

2. Alkaline trachyte, mean of 21 analyses. Osann-Rosenbusch, *Gesteinslehre*, 1923, p. 378.

3. Phonolite, mean of 25 analyses. Daly, *op. cit.*, p. 24.

4. Trachyandesite, mean of 19 analyses. Osann-Rosenbusch, *op. cit.*, p. 416.

composition of the average andesite and basalt shows that the essential differences between the two types depend on the proportion of felsic to mafic constituents. Andesite is a dofelsic, and basalt a mafelsic type; i.e. in andesite more than five-eighths of the rock consists of felsic constituents, whereas in basalt there should be approximate equality between the two mineral groups.

Andesites may be concisely defined as those lavas in which plagioclase felspar is the predominating constituent. Alkali-felspar may occur in subordinate amount; and quartz, although not usually visible, is nevertheless latent or occult in the cryptocrystalline or glassy parts of the groundmass. As mentioned elsewhere (p. 109), computation of the standard mineral composition of the average andesite gives 15 per cent. of free silica. This figure is probably too high, for a critical revision of the analyses which make up the average would certainly result in the rejection of some as not typical andesites. It may be concluded, however, that andesite is an over-saturated lava, which may contain excess silica up to a maximum of about 10 per cent. The ferro-magnesian minerals may be biotite, hornblende, augite, enstatite, or hypersthene, giving rise to varieties named accordingly as biotite-andesite, etc. In some of the more basic types sporadic olivine may be present, which must be regarded in this case as a reactional mineral (p. 78). Andesites are usually porphyritic with felspars and the ferro-magnesian minerals as phenocrysts. The groundmass shows trachytic or hyalopilitic texture. Glassy rocks of andesitic composition are known, which may be called *vitro-* or *hyalo-andesite* according to the amount of glass present. Some pitchstones are of andesitic composition.[1]

The alteration of andesites by hot carbonated waters, leading to the formation of epidote, calcite, chlorite, and serpentine from the pre-existing minerals, produces the greenish or greyish decomposed rocks known as *propylite*. As propylitisation is a very widespread phenomenon in mineralised andesite regions, the term propylite, which tends to become obsolete, should be retained for the distinctive alteration products.

Andesites are the characteristic rocks of volcanoes which arise on or near, or in connection with, the great fold-mountain chains of the earth. Their name indicates their occurrence in the Andes, but unfortunately the Andean andesites are the least well-known rocks of the group. The recent and present-day volcanoes which girdle the Pacific erupt andesites, especially hypersthene-bearing varieties. The Hungarian, Balkan, Caucasian, and Persian andesite regions are of Kainozoic age, and were erupted in connection with the building of the Alpine-Himalayan chains. Of more ancient andesite lava fields there are numerous examples. In Scotland the Old Red Sandstone lavas are mainly andesitic (Ochil Hills, Cheviots, Carrick Hills, Lorne), and olivine-bearing types are common. The summit of Ben Nevis consists of a cake of hornblende-andesite lava of Lower Old Red Sandstone age. The Lower Palæozoic lavas of Wales and the Lake District (Snowdon, Borrowdale, etc.), are also largely

---

[1] Anderson and Radley, " Pitchstones of Mull and their Genesis,' *Q.J.G.S.*, lxxi, 1917, pp. 205-17.

of andesitic composition, and are associated with abundant rhyolites.

The term *basalt* was used by Pliny, and is said to be derived from an Ethiopian word signifying a black, iron-bearing stone. It was first applied as a definite petrographic name by Agricola to certain Saxon rocks. In modern usage basalts may be defined as mafelsic lavas in which plagioclase felspar and the group of mafic minerals occur in approximately equal amounts. The mafic minerals consist of augite, olivine, and iron oxides; hypersthene, hornblende, or biotite, only occur in exceptional cases. The plagioclase, while commonly labradorite, may range from oligoclase to anorthite, and there may be a small residue of alkali-felspar and quartz. The term *basalt* is applied to the simple mixture, labradorite, augite, iron oxides; *olivine-basalt*, when olivine is present in notable amount.

Most basalts are slightly undersaturated rocks; and when the deficiency in silica, or the abundance of soda and potash, are sufficient to bring in a small amount of nepheline, analcite, or potash-felspar, we have the abundant class of *alkali-basalts* or *trachybasalts* (formerly trachydolerite).[1] An especially noteworthy Scottish member of this class is the rock called *mugearite* (Mugeary, Skye) by Harker.[2] Mugearite consists of oligoclase with a notable amount of orthoclase, with olivine, augite, and iron oxides, as the mafic minerals. It passes over to the trachyandesites by increase of the felspars. It has been found abundantly in the Scottish plateau lavas of Carboniferous age, as well as in the Kainozoic fields of the Western Isles.

Spilite is a soda-rich basaltic type in which albite or albite-oligoclase is the predominating felspar. These rocks are usually much altered with the formation of carbonates, and the pyroxene in most cases has completely gone over to chlorite or serpentine. Pillow structures (p. 37) are more perfectly developed in spilites than in any other igneous rock.[3]

On the other hand, certain basaltic magmas are saturated, or slightly oversaturated, with silica. The silica may appear

---

[1] Report of Committee on Petrographic Nomenclature, *Min. Mag.* xix, 1921, p. 144.

[2] *Tert. Ign. Rocks of Skye*, 1904, p. 265.

[3] A. K. Wells, *Geol. Mag.*, lx, 1923, pp. 62-74.

in crystalline form (in *quartz-basalt*), or it may remain occult in the groundmass (*bandaite*). Many of these rocks are on the borderline between andesite and basalt. In fact, bandaite, which is a labradorite-pyroxene lava with excess silica, has varieties which pass over into the andesite class. Study of the chemical and mineral characters of the rocks known as *tholeiite*, which have been described from the Kainozoic igneous fields of the west of Scotland, shows that the term tholeiite, originally applied by Rosenbusch to olivine-free and olivine-poor plagioclase-augite rocks with intersertal texture, can also be usefully given to oversaturated basalts rich in lime, in which highly-calcic plagioclase felspars appear, mostly as phenocrysts. In Scottish petrology the term is applied to intrusive rocks of this character which possess intersertal texture.[1] Basaltic glass, or *tachylyte*, is comparatively rare, and is mainly found on the chilled margins of basaltic intrusions.

Basaltic rocks in which nepheline, analcite, or leucite become abundant and essential constituents, may also be dealt with here. Types free from olivine, and containing both plagioclase and a felspathoid, are known as *tephrite ;* with olivine, the term *basanite* is used. When the felspar becomes very subordinate, or is absent, so that the felspathoid predominates, types free from olivine are called *nephelinite* or *leucitite*, and those rich in olivine are known as *nepheline-, analcite-,* or *leucite-basalt*, according to the particular felspathoid present.

Some basaltic types are so rich in olivine, augite, and iron ores, that they become ultrabasic in composition, and need to be distinguished from basalts proper. The felspars are either absent or very subordinate in amount. When augite is the predominant mafic mineral the term *ankaramite* (Ankaramy, Madagascar) is applied ;[2] for types rich in olivine the name *oceanite* has recently been proposed by Lacroix.[3] Glass-rich rocks corresponding to ankaramite are known as *augitite*, and those corresponding to oceanite as *limburgite*.[4]

[1] *Mull Memoir*, 1924, pp 280, 284, 370-2.
[2] Lacroix, *Comptes Rendus*, Paris, tom. 163, 1916, p. 182.
[3] Lacroix, *Mineral. de Madagascar*, tom. 3, 1923, pp. 49-50.
[4] Osann-Rosenbusch, *Gesteinslehre*, 1923, p. 491.

Common basalt occurs in great confluent sheets forming lava floods which, in places, cover areas of the order of hundreds of thousands of square miles, with an average thickness of at least half a mile. The Deccan of Peninsular India; the Columbia and Snake River plains of Western America; the great basalt plateau of which part of Greenland, the basement of Iceland, the Faröe Islands, Skye, Mull, and Antrim are the remnants; and the Parana basalts of South America; are examples of these enormous fields.

Olivine-bearing basalts are the chief products of the great shield volcanoes which rise from the ocean depths (Hawaii), and those that form the superstructure of Iceland, and the bulk of many of the smaller oceanic islands. In the latter the olivine-basalts are associated with abundant trachybasalts, and subordinate trachyandesites and trachytes. Along with these, ankaramites and oceanites occur as differentiates in the ultrabasic direction; while nepheline-basalts, tephrites, and phonolites, may occur as small, localised differentiates in an extreme alkalic direction. Of ancient basalt fields of similar character may be mentioned that of the Carboniferous in Scotland, in which basalts of slightly alkalic character are associated with mugearites, trachytes, and phonolites.

Spilites seem to be the characteristic lavas of geosynclinal deposition, and are extruded on, or just beneath, the ocean floor in regions of rapid sedimentation. As these accumulations are later uplifted and folded to form portions of mountain chains, the spilites and their associates have frequently suffered a low-grade metamorphism which has converted them into the well-known " green rocks," ophites, and serpentines, of fold-mountain regions.

The chemical composition of andesites and basalts is illustrated by the analyses of Table IX.

TABLE IX

	1	2	3	4	5	6
SiO_2 . . .	60·8	59·3	48·8	48·8	39·2	45·6
Al_2O_3 . . .	17·3	16·6	15·8	16·0	12·6	8·3
Fe_2O_3 . . .	2·9	3·1	5·4	4·3	5·9	2·3
FeO . . .	2·5	3·5	6·3	6·3	7·4	10·2
MgO . . .	2·5	3·4	6·0	5·4	11·6	21·7
CaO . . .	5·5	6·3	8·9	8·2	12·9	7·5
Na_2O . . .	4·0	3·6	3·2	3·9	3·6	1·3
K_2O . . .	2·4	1·9	1·6	2·8	1·7	·4
H_2O . . .	1·2	1·3	1·8	1·5	1·9	·6
TiO_2 . . .	·6	·7	1·4	1·9	2·3	1·7
P_2O_5 . . .	·2	·2	·5	·7	·8	·3
MnO . . .	·1	·1	·3	·2	·1	·1

Summation, 100.

1. Biotite- and hornblende-andesites, mean of 18 analyses. Osann-Rosenbusch, *Gesteinslehre*, 1923, p. 405.

2. Hypersthene- and augite-andesites, mean of 20 analyses. *Ibid.*, p. 409.

3. Basalt, mean of 161 analyses. Daly, *Igneous Rocks and Their Origin*, 1914, p. 27. See also average analysis of plateau basalt (p. 54).

4. Trachybasalt, mean of 28 analyses. Osann-Rosenbusch, *op. cit.*, p. 460.

5. Nepheline-basalt, mean of 21 analyses. *Ibid.*, p. 486.

6. Oceanite (ultrabasic olivine-rich " basalt "), mean of 10 analyses from Madagascar, Hawaii, etc.

CHAPTER VII

THE DISTRIBUTION OF IGNEOUS ROCKS IN SPACE AND TIME

CONSANGUINITY—The term *consanguinity* (Iddings) [1] is used to indicate the fact that certain groups of igneous rocks, the members of which are associated in space and time, possess a community of character or family likeness which is expressed in their chemical, mineralogical, textural, and geological features. While in chemical composition consanguineous series or suites may range from acid to ultrabasic types, some mineral and chemical characters are *constant*, i.e. are common to practically all members ; while other characters are *serial*, that is to say, they show a regular variation throughout the series. Thus, in some suites, a constant character is oversaturation with silica, which causes free silica to appear in quite basic members. A serial character may be afforded by the regular variation of the alkalies, or of ferrous iron oxide and magnesia throughout the suite. Some series may be characterised throughout by a peculiar mineralogical feature, such as the occurrence of anorthoclase, as in certain Norwegian, East African, and Antarctic suites. Consanguinity in an igneous series leads to the hypothesis that the assemblage has been derived by some process of differentiation (Chap. VIII) from a common initial magma, or from a number of closely related magmas.

THE DIAGRAMMATIC REPRESENTATION OF IGNEOUS ROCK SERIES [2]—The chemical and mineral relationships obtaining

---

[1] *Bull. Phil. Soc., Washington*, xii, 1892, pp. 128-30. As far back as 1880 J. W. Judd remarked on the " family likeness " between certain groups of lavas. *Volcanoes*, 1880, p. 202.
[2] A. Harker, *Nat. Hist. Ign. Rocks*, 1909, pp. 118-32; A. Holmes, *Petr. Methods and Calc.*, 1921, Chap. XI.

in a consanguineous series can be exhibited by suitable graphs, and may thus be rendered in forms which permit rapid visualisation of their characters. Diagrams of this kind are called *variation-diagrams*. The diagram which is most frequently used is one in which silica is plotted against the other oxides in such a way that an analysis is represented by a series of points on a vertical line. By joining up the points for each oxide in a series of analyses, curves are obtained which show

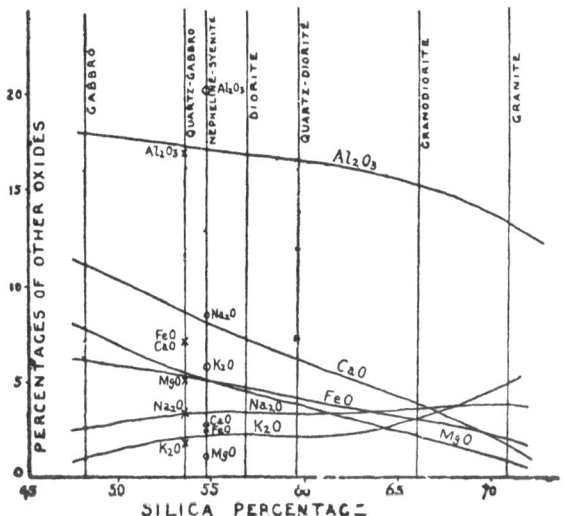

FIG. 44.—SILICA VARIATION DIAGRAM OF THE GRANITE-DIORITE
SERIES.
(Text, p. 134).

graphically the variation of each constituent with regard to silica. As the amount of silica constitutes a rough index of the stage of differentiation a rock has reached, the curves collectively present a picture of the course of differentiation which has produced the rock series under examination, as well as a succinct statement of its chemical characters.

Fig. 44 illustrates a typical variation diagram, obtained by plotting analyses 2, 3, 4, and 5 in Table IV, and 2 in Table VI.

Together these analyses represent a very common type of igneous rock series which ranges from granite to gabbro through the intermediate stages of tonalite, quartz-diorite, and diorite. The curves are fairly regular and require very little smoothing, indicating that the analyses plotted have real serial relations. The crosses indicate the points for the analysis of quartz-gabbro (Table VI, 1). The aberrant position of these points shows that this rock is alien to the series ; and if included it would seriously disturb the regularity of the curves. Nepheline-syenite (Table V, 4) the analysis of which is represented by circles, is clearly an even more discordant rock in this series than quartz-gabbro.

The curves for Al_2O_3, CaO, FeO, and MgO, all bend downwards to the right of the diagram, indicating a decrease in these constituents as the acid end of the series is approached. These constituents are said to vary sympathetically. The soda and potash curves, however, bend upward to the right, indicating an increase in these constituents in the acid members of the series. They vary sympathetically with each other, but are in antipathetic relation to Al_2O_3, CaO, FeO, and MgO.

It has been emphasised (p. 106) that variation in an igneous rock series takes place mainly in respect to the relative proportions of the felsic and mafic groups of minerals. The silica percentage provides a rough index of this variation in many cases, but it completely fails in some series, especially those of alkali-rich rocks. The proportion of salic minerals, calculated from the norm (p. 103), gives, on the whole, a much better index of variation. A variation-diagram of the average granite-gabbro series, with the percentage of salic minerals substituted for the silica percentage is given in Fig. 45. The curves follow much the same courses as in Fig. 44, but are more regular, although the improvement is not so apparent in this series as in many others.

In this diagram is also shown a curve of the *silica number*, which represents the excess or defect of silica in molecular values, in respect to the amount which is required just to saturate the rock. It is obtained from the calculation of the norm. Free quartz computed in molecules gives the excess silica, while deficits are calculated from the amounts of olivine and felspathoids in the norm. In the series illustrated by

Fig. 45, only the gabbro gives a deficit of silica. The curve cuts the base-line, i.e. the silica number is zero, at the point representing 65 per cent. of felsic minerals. A perpendicular erected at this point may be called the *saturation line*. The points at which it cuts the curves gives the approximate composition of the just-saturated rock of the series under

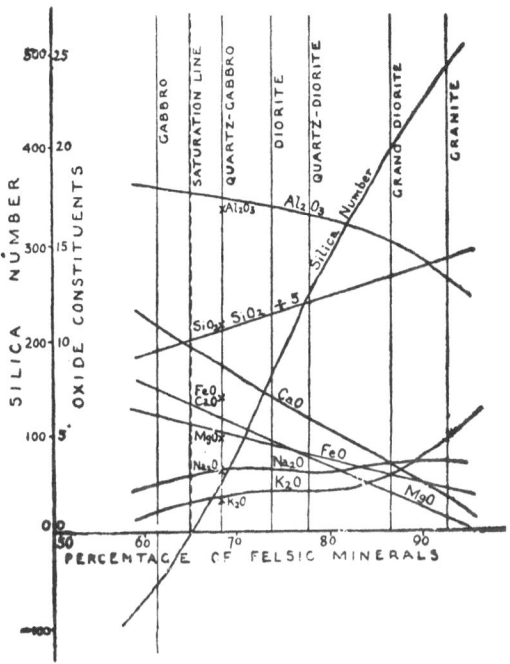

FIG. 45.—VARIATION DIAGRAM OF THE GRANITE-DIORITE SERIES.
(For explanation see text, p. 134.)

consideration, which may be called the *saturation composition*. The position of the saturation line, and the saturation composition, vary in different series (Figs. 45, 46), and are thus of diagnostic value. For a discussion of other types of variation diagram the works of Harker and Holmes cited above should be consulted.

KINDREDS OF IGNEOUS ROCKS—Groups, series, or suites of igneous rocks which show consanguineous chemical and mineral characters, and which appear to be genetically related, may be called *kindreds*. Several terms have been used in closely similar senses, as series, suite, tribe, clan, branch, stem (German *stamm*). The relationship implied by consanguinity may be of different degrees of closeness. The term kindred may be used for the widest kind of grouping in which the consanguinity between the rocks is of the most tenuous kind. Within a kindred may be found *tribes*, the rocks of which show a closer relationship ; and within tribes, *clans*, with a very high degree of consanguinity. Kindreds, tribes, and clans, cut across formal classifications ; and certain abundant rock types, such as basalt, andesite, and granite, may recur in several different groupings. Although it may be composed of a wide diversity of rock types, a kindred is marked first of all by a certain community of chemical and mineral characters, which may be exhibited in constant or serial relations. It may be further characterised by the different bulk-development of its component rock types. Thus, in certain kindreds andesites are the most abundant rock types, and basalt is a subordinate member ; whilst in others, basalts predominate greatly over andesites. The degree of differentiation is another factor whereby kindreds may be distinguished. Some kindreds consist of huge masses of uniform rocks (e.g. the basalt floods) ; others contain a great diversity of types within a small compass. Furthermore, there are differences among kindreds in regard to the relative development of plutonic, hypabyssal, and volcanic facies, at least in the accessible crust. This may, of course, be due to accidents of erosion and exposure, but it is believed that there are also intrinsic differences in this respect. Finally, kindreds are distinguishable by their extension in space and time (petrographic provinces and periods) and by their varying relations to geotectonic processes. These topics will be dealt with in later sections.

The progress of research has not yet advanced far enough for all the different kindreds to have been discriminated, much less for tribes and clans. The subject is still being actively investigated. We shall proceed to a brief survey of the kindreds which have already been proposed.

A very wide and vague, but nevertheless valid, grouping of igneous rocks, is that into *alkalic* and *calcic* (= subalkalic, calc-alkalic), which has been much exploited in the past.   It is probable that both terms cover several distinct kindreds, and that some kindreds pass over the gradational boundary between the two groups.   The terms alkalic and calcic, nevertheless express a real tendency in igneous rocks for the separation of two broadly-contrasted assemblages, which have different chemical, mineral, geographical, and geotectonic relations.   The mineralogical contrast may be expressed as below :—[1]

Alkalic.	Calcic.
Alkali-felspar abundant in the felsic and intermediate rock-types, and even in some mafic types.	Alkali-felspar not common except in the more silicic rock types.
Soda-lime felspars not common except in the more mafic types.	Soda-lime felspars abundant throughout the range.
Felspathoids often occur.	Felspathoids absent.
Quartz appears only in the more silicic rock types.	Quartz occurs throughout most of the range of types, either crystallised or occult.
Pyroxenes and amphiboles of alkali-rich varieties. Rhombic pyroxene absent.	Ordinary augite, hornblende, and rhombic pyroxenes present.
Micas and garnets common.	Micas not common except in the more silicic rock types. Garnets very rare.

Chemically, alkalic rocks are characterised by high percentages of alkalies in relation to silica and alumina.   In calcic rocks the ratio of alkalies to silica and alumina is not so high, and constituents such as lime and the ferro-magnesian oxides are relatively more abundant.   In alkalic kindreds saturation occurs at a relatively high silica percentage and felsic/mafic ratio ; whereas in calcic kindreds the saturation composition falls at a low silica percentage, and a comparatively small felsic/mafic ratio (p. 105).

It remains now briefly to indicate the characters of some other kindreds.   In 1911 H. Dewey and Sir J. S. Flett [2]

[1] A. Harker, *Nat. Hist. Ign. Rocks*, 1909, p. 91.
[2] *Geol. Mag.*, 1911, pp. 202-9, 241-8.

established the *spilitic* suite, in which the commonest types are spilite and diabase, with keratophyre, soda-granite, and picrite, as less frequent members. In some regions gabbros and serpentines are suspected to be the plutonic representatives of the suite. The essential chemical character is the great relative abundance of soda, which results in albite being the most typical felspar of the kindred. Albitisation is common in the more basic members. Spilitic lavas are very commonly pillow-form, and are frequently associated with radiolarian cherts and other marine sediments.

F. von Wolff [1] has proposed an " Arctic " suite to include the vast floods of basalt which appear in Peninsular India, Siberia, and the North Atlantic or Thulean region. The overwhelmingly predominant rock type in this kindred is a slightly over-saturated basalt which, from its mode of occurrence, is called *plateau-basalt*. Very subordinate rhyolites and trachytes have been recorded as occurring along with these basalt floods (Deccan, Siberia, Iceland), but very little is as yet known regarding these rocks. The Arctic kindred is, therefore, characterised by extreme uniformity of chemical character, and by its occurrence mainly in volcanic form.

A kindred which probably has close affinities to the plateau-basalts is that in which the characteristic rock type is quartz-gabbro or quartz-dolerite.[2] These rocks are very widespread (Scotland, Eastern North America, Guiana, Argentina and Uruguay, South Africa, Antarctica, and West Australia) as thick dykes and massive sills, and in some cases, as huge lopoliths (Duluth, Insizwa, etc., see p. 20). In the larger masses a long series of differentiation products is found, ranging from soda-granite, through intermediate types, to the dominant quartz-gabbro, and ending with olivine-gabbros and ultrabasic rocks.

The common granodiorite–andesite kindred is the one which has been formulated as the typical example of a calcic igneous series. Its plutonic members range from peridotite, olivine-gabbro, to gabbro ; and thence, through diorite, quartz-diorite, and granodiorite, to granite. The commonly

---

[1] *Vulkanismus*, 1914, p. 153.
[2] G. W. Tyrrell, "Geology and Petrology of the Kilsyth-Croy District," *Geol. Mag.*, 1909, p. 362.

associated volcanic rocks are andesite, dacite, and rhyolite. The variation-diagrams, Figs. 44, 45, exhibit the chemical characters of this kindred.

V. M. Goldschmidt[1] has also distinguished a " mica-diorite " kindred, characterised by the early and abundant separation of biotite in a differentiation series of the grano-diorite–andesite type. The early withdrawal of potash in biotite leads to the production of soda-rich types in later stages, so that one of the end-products is an oligoclase-rich granite (trondhjemite). The principal rock types are biotite-norite, biotite-diorite, and trondhjemite. This series should, perhaps, be regarded as a tribe or clan of the granodiorite–andesite kindred.

An important kindred is that which is characterised by the presence of anorthosite and charnockite (hypersthene-granite). Anorthosite may occur in other kindreds through local differentiation, but in this kindred it occurs as some of the largest igneous rock bodies known. The term char-nockite series denotes a series beginning with norite (or even pyroxenite) and ending with pyroxene-granite. One of the distinguishing features of the kindred is its poverty in water-rich, or water-formed, minerals, such as biotite and horn-blende, although these minerals are not absent from some members. Typical " dry " minerals, such as the pyroxenes, persist even into the extreme acid end-members of the series. Rosenbusch first hinted that the anorthosite–charnockite kindred might be separable from the normal calc-alkalic series.[2] Goldschmidt,[3] and J. H. L. Vogt,[4] however, have more fully described its characters. From South Norway, Goldschmidt has described the series norite—Jotun-norite (potassic)—mangerite (pyroxenic monzonite)—hypersthene-syenite — hypersthene-granite — diopside-granite — aegirine-granite, in connection with which anorthosites and pyroxen-ites are formed as gravity differentiates from the more basic

[1] *Stammestypen der Eruptivgesteine. Vid.-selsk. Skr* 1, *Math.-Nat. Kl. Kristiania*, 1922, No. 10, p. 6.

[2] *Elem. der Gest.*, 2nd ed., 1901, p. 163.

[3] *Op. cit. supra*, p. 8.

[4] "The Physical Chemistry of the Magmatic Differentiation of Igneous Rocks," *Vid.-selsk. Skr.* 1, *Math.-Nat. Kl. Kristiania*, 1924, No. 15, pp. 52-98.

magmas.   This " mangerite stem " appears to be a tribe of the anorthosite–charnockite kindred.

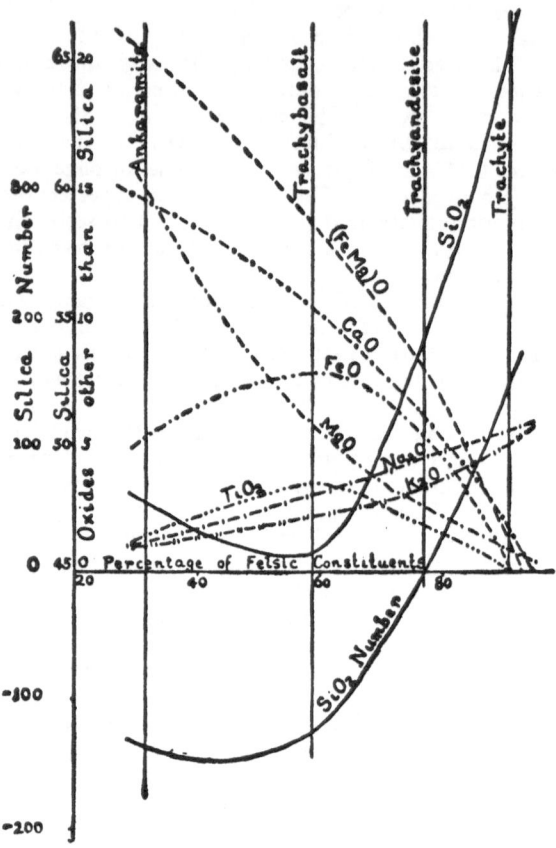

FIG. 46.—VARIATION DIAGRAM OF THE LAVAS OF JAN MAYEN.
(For explanation see text, p. 141.)

A common kindred is that which Holmes has designated as trachydoleritic (trachybasaltic).[1]   It occurs most frequently

[1] *Min. Mag.*, xviii, 1918, p. 220.

in oceanic islands and in regions of tension and subsidence. The most abundant rock type is a slightly undersaturated basalt, which differentiates in the one direction through trachybasalts and trachyandesites, to trachytes and soda-rhyolites ; and in the other direction to the olivine-rich basalts ankaramite and oceanite. Occasionally more richly alkaline magma appears through local differentiation, giving rise to nepheline-basalt, tephrite, and phonolite. The kindred is mainly volcanic, but plutonic types may occur, and consist of essexite, teschenite, crinanite, soda-syenite, and infrequently more richly alkaline rocks. A typical example of this kindred is given by the volcanic rocks of Jan Mayen (Fig. 46). In this series the saturation composition gives 54·5 per cent. of silica, and has about 76·5 per cent. of felsic constituents. For the rocks of Ascension Island the corresponding figures are 54 per cent. silica and 70·5 per cent. felsic constituents.

Richly alkaline rocks, such as the ijolite–melteigite series of Norway, Finland, Arkansas, and the Transvaal, with the associated nepheline-syenites, constitute a definite kindred, which, however, is of rare occurrence. It is mostly of plutonic habit, and probably originates in more than one way (p. 167).

Finally, there may, perhaps, be distinguished a kindred characterised by rocks relatively rich in potash, in which such minerals as leucite, potash felspar, and biotite, are prominent constituents. Niggli has designated this kindred as *Mediterranean,*[1] and has included therein magmas of quartz-syenite, syenite, monzonite, and shonkinite composition. Potassic rocks occur quite frequently in association with the trachybasaltic and other alkalic kindreds. It is possible that several kindreds are included in the richly sodic and potassic rocks, and much more research is needed to systematise them.

PETROGRAPHIC PROVINCES AND PERIODS [2]—A *petrographic province* is the geographical extension of a kindred, and a *petrographic period* is likewise its extension in time. The

---

[1] *Lehrbuch d. Min.*, 1921, p. 493.
[2] A. Harker, *Nat. Hist. Ign. Rocks*, 1909, Chap. IV ; M. Stark, *Petrographische Provinzen, Fortschr. d. Min. Krist. u. Petr.*, 4, 1914, pp. 251-336.

first of these terms was proposed by J. W. Judd [1] in the following words: "There are distinct petrographical provinces within which the rocks erupted *during any particular geological period* present certain well-marked peculiarities in mineralogical composition and microscopical structure, serving at once to distinguish them from the rocks belonging to the same general group, which were simultaneously erupted in other petrographical provinces." A. Harker's definition [2] is that a petrographic province "is a more or less clearly-defined tract within which the igneous rocks *belonging to a given period of igneous activity*, present a certain community of petrographical character traceable throughout all their diversity, or at least obscured only in some of the more extreme members of the assemblage." In both these definitions the phrase italicised insists on a time limitation. Many objections to the idea of petrographical provinces have been based on the mistaken view that the term applied to all the igneous rocks within a given region regardless of their ages.

The boundaries of petrographical provinces and periods are vague and ill-defined because the true spatial and temporal associations of a kindred are with a certain geological environment. Hence the modern tendency is to place much more emphasis on the nature of the kindred itself, and on its connection with a definite tectonic process, than on its mere geographical extension. For this reason the geographical names which have been applied alike to petrographic provinces and to kindreds, such as Atlantic for the alkalic kindreds, Pacific for the calc-alkalic kindreds, and the later terms, such as Arctic and Mediterranean, are to be deprecated as useless and misleading. [3] It is nevertheless true that, for igneous rocks erupted within a given period of magmatic activity, large areas of the earth's surface can be mapped out into more or less definite provinces in which different kindreds hold sway. The continued use of the term petrographic province is therefore justified. Similarly, the

---

[1] "On the Gabbros, Dolerites, and Basalts of Tertiary Age in Scotland," *Quart. Journ. Geol. Soc.*, xlii, 1886, p. 54.
[2] "Some Aspects of Modern Petrology." Pres. Address to Geol Sect. Brit. Assoc. for Adv. of Science, Portsmouth, 1911, p. 3. Reprint.
[3] See J. W. Gregory, "Structural and Petrographic Classification of Coast Types," *Scientia*, 1911, pp. 37-63.

term petrographic period may be used to indicate that petro-graphic provinces have a more or less definite extension in time as well as in space.

IGNEOUS ACTION AND EARTH MOVEMENTS—The study of geological history shows that igneous action has not been continuous in any one region. Periods of activity have alternated with periods of quiescence ; and it is found that the extrusion and intrusion of igneous rocks coincide more or less closely with the periods in which earth movements reach their maximum intensity. Igneous action has, in general, taken place at those times (petrographic periods), and over those regions (petrographic provinces), in which epeirogenic (continent-making) and orogenic (mountain-making) move-ments have been in progress. At any one period of earth history, therefore, the crust may be divided into regions of tension and regions of compression. Not only does the period and locus of igneous action correspond with earth movement, but the assemblages (kindreds) of igneous rocks produced vary with the kind of earth movement, and with the geological environment. Some kindreds are associated with the slow vertical movements of large blocks of the earth's crust, and with their fracturing and faulting (epeirogeny) ; others are associated with the relatively rapid and short lateral movements which produce folding and overthrusting, and which act on long narrow strips of the crust forming fold-mountain ranges (orogeny). It has been suggested that the broad class of alkaline rocks is connected with the former type of earth movement, and the calc-alkali class with the latter ; but the vague character of the subdivision led to the discovery of numerous apparent exceptions to the rule. On the whole, alkalic kindreds do tend to be associated with epeirogenic, and calc-alkali kindreds with orogenic move-ments of the earth's crust.

The granodiorite–andesite kindred is typically associated with fold-mountain ranges, and appear to be in genetic con-nection with the movements which have produced them. Thus the more deeply-worn central parts of the fold-mountain chains that girdle the Pacific Ocean, and those of the Alpine-Himalayan system, often expose cores of the plutonic rocks of this kindred ; whilst the volcanoes which exist on or near the mountain chains erupt mainly andesitic lavas. At the present day many of the active volcanoes in the above-mentioned regions are erupting a very characteristic hypersthene-andesite lava.

Deep sections of old mountain chains, such as the Caledonian of Norway, as also many Archæan regions, exhibit petrographic provinces of the anorthosite–charnockite type. This kindred is characterised in general by the absence of water-rich or water-formed minerals, and by coarse granular textures. These features are probably due to deep-seated intrusion at very high temperatures into " dry " rocks, a geological environment which would explain the rarity of hypabyssal, and the absence of volcanic rocks, belonging to this kindred.[1]

According to Dewey and Flett,[2] the spilitic kindred is characteristic of regions which have undergone a long-continued gentle subsidence, with few or slight upward movements, and no important folding. The frequent association of spilitic rocks with particular types of sediment, and their location in fold-mountain regions, where they appear as " green rocks " of low-grade metamorphic types, establish them as the characteristic igneous rocks of geosynclinal regions. These zones of thick sedimentation are later uplifted and folded to form parts of mountain ranges, their igneous rocks along with them suffering the same degree of metamorphism.[3]

The plateau basalt kindred seems to be associated with the major earth movements by which oceans have been brought into being. Thus the Thulean plateau basalts appear to be connected with the crustal inbreak that initiated the North Atlantic ; and the Deccan plateau basalts with the similar inbreak by which the northern part of the Indian Ocean was formed.

The closely-associated quartz-gabbro kindred is in some way connected with great regions of crustal tension, as is shown by the frequent occurrence of parallel systems of

---

[1] See papers by Goldschmidt and Vogt cited on p. 139.

[2] *Geol. Mag.*, 1911, p. 246.

[3] The view may perhaps be hazarded that the abundance of soda in spilitic rocks is in some way connected with the richness of geosynclinal sediments in connate sea salts. See H. I. Jensen, *Proc. Linn. Soc., N.S.W.*, 33, 1908, p. 518.

massive dykes. But it is not yet clear whether this kindred is an after-effect of previous mountain-making movements, or whether it represents the continental equivalent of plateau-basalts.

The trachybasalt kindred appears to be connected with a wide range of geological environment. It occurs super-posed on plateau-basalts as in Iceland and Kerguelen ; in islands which rise abruptly from oceanic depths, as in Ascension, Samoa, etc. ; in major regions of tension, as in the rift valleys of East Africa ; and in regions which have undergone adjustment by block faulting after orogeny, such as the extra-Alpine petrographic provinces from Auvergne to Bohemia.

Similarly, the highly-sodic rocks rich in nepheline, and the potassic leucite rocks, are found in several kinds of geological environment, and much more work needs to be done before their tectonic relations can be elucidated. In fact, the whole subject of the connection of igneous rocks with tectonic and other geological features is still in the stage of speculation, and established results are still to seek.

TIME SEQUENCES IN IGNEOUS ROCKS—It has been shown that the igneous rocks erupted at a particular centre during a given period are chemically and mineralogically related in such a way as to suggest a common derivation. The order in which the various types have been erupted is also of significance for the understanding of the processes which have gone on in the magma reservoir. No definite order of succession, however, is discernible if all the igneous rocks of a province are arranged chronologically without regard to the phase or category of igneous action which they represent. But if plutonic, hypabyssal, and volcanic rocks, are considered separately, definite sequences of eruption often become clear, which are different in each phase because plutonic, hypabyssal, and volcanic rocks come into being under different geological conditions. In plutonic rocks the general sequence of intrusion is from basic to acid. Thus, in the Kainozoic province of Skye, peridotites were intruded first, then gabbros, and finally granite. The Oslo (Kristiania) province of Devonian age in Norway shows the following sequence :—

1. Essexites (with ultrabasic pyroxenic facies).

10

2. Larvikite and lardalite (comparatively-basic soda-syenites and nepheline-syenites rich in pyroxenes, olivine, and
iron-ore).

3. (a) Nordmarkite (quartz-bearing soda-syenite).
   (b) Soda-granite.
   (c) Biotite-granite.
   (d) Rapakiwi-granite (porphyritic granite rich in orthoclase).

This order holds for plutonic series of widely-different characters from several kindreds.   The sequence is parallel with
that due to a common type of differentiation (p. 156), whereby
basic rocks rich in the early-crystallised magnesia-rich minerals are separated from more acid and alkaline rocks in which
the minerals of late crystallisation are concentrated.   The
latter remain liquid longest, and hence commonly have intrusive relations to the already solid basic rocks.   Also
the more acid and alkaline magma types can only appear in
bulk at a late stage in differentiation.

The actual time order of a series of volcanic rocks is more
easily determined than that of a plutonic series, because lavas
are often piled one upon the other like a stratigraphical sequence, and the law of superposition thus becomes applicable.
It often happens, however, that no regularities are to be dis-
covered in a sequence, as significant relations may be masked
by the overlapping of the products of two or more adjacent
volcanoes, or through other accidental circumstances attending the eruptions.

It has been claimed that, in simple cases, the order of
eruption is that of increasing divergence from an initial
type.   This may be illustrated by the succession in the Late
Kainozoic lavas of the Eureka district, Nevada : 1, horn-
blende-dacite ; 2, hornblende-mica-dacite ; 3, dacite ; 4,
rhyolite ; 5, pyroxene-andesite ; 6, basalt.   Here the rocks
may be arranged in two interdigitating sequences, 1, 2, 3,
and 4 making up a series of increasing acidity ; and 1, 5,
and 6 representing a series of increasing basicity.   The initial
type must obviously be a rock of intermediate acidity.

Study of the many definite sequences of volcanic rocks
given by Daly,[1] however, shows that the question is one of

---

[1] *Igneous Rocks and Their Origin*, 1914, Appendix B, p. 469 *et seq*.

considerable difficulty. A long succession is often punctuated, so to speak, by the appearance of a constant type, especially basalt and rhyolite. As Harker has shown, this may be due to fresh accessions of magma bringing about repetition of the same sequence. Thus, in the volcanic succession of the Berkeley Hills, near San Francisco, writing a for andesite, b for basalt, *r* for rhyolite tuff, the sequences are as follows :—

Lower Berkeleyan formation .	.	a, b, *r* ; a, b, *r*.
Upper „ „	.	a, b, *r* ; a, b.
Campan formation	. .	a, b, r, b, *r*.

Many successions begin with rhyolite or some other acid or intermediate lava, and end with basalt, the intervening members being of somewhat variable acidity. The Middle Kainozoic lavas of Victoria provide an excellent example.[1] The succession is 1, solvsbergite ; 2, alkaline trachyte ; 3, anorthoclase-basalt; 4, olivine-trachyte ; 5, olivine-anorthoclase-basalt ; 6, limburgite ; 7, basalt. Similarly, in the Mont Dore region of Auvergne [2] the sequence from Mid-Pliocene to Recent is as follows : 1, phonolite ; 2, rhyolite ; 3, basalt ; 4, andesite and basalt ; 5, acid tuffs ; 6, acid andesite and trachyte ; 7, augite-andesite and tephrite ; 8, phonolite ; 9, plateau-basalts ; 10, basalt. The genetic meaning of these sequences is not yet clear, and much more work is needed for their elucidation. It is probable that in the majority of cases, as stated above, the significant order of eruption is obscured by adventitious circumstances.

In the case of hypabyssal intrusions it is very often difficult to make out any order of succession, for the reason that the various types are only exceptionally seen in contact with one another. But where the order of intrusion has been elucidated, it has been found that, for minor intrusions related to a particular focus of igneous activity, there is frequently a general sequence from acid to basic types, just as in many volcanic sequences, and the reverse of the common plutonic sequence.

[1] *Igneous Rocks and Their Origin*, 1914, Appendix, B, p. 484.
[2] *Ibid.*, p. 480.

CHAPTER VIII

ORIGIN OF IGNEOUS ROCKS

VARIATIONS IN IGNEOUS ROCKS—It has been shown that the composition of igneous rocks as a whole varies widely within certain definite limits (p. 46). This diversity may be shown in a number of related but separate intrusions or extrusions; or it may appear within a single rock mass. Reasons have been given for the assumption that there are only a few, perhaps only two, primary magmas (p. 53). Two problems are thus involved: namely, that of the derivation of the present diversity of igneous rock types from simple initial magmas; and that of the ultimate origin of the primary magmas themselves. With regard to the latter, only speculation can, as yet, be offered, and it is not proposed to deal with this question here; but the problem of the immediate origin of igneous rocks has already been more or less successfully tackled. The present variations of igneous rocks may be ascribed to two causes: differentiation, and assimilation (or syntexis). *Differentiation* may be defined as the process whereby a magma, originally homogeneous, splits up into contrasted parts, which may form separate bodies of rock, or may remain within the boundaries of a single unitary mass. *Assimilation* is the process whereby foreign rock material, either in liquid or solid form, is incorporated within a magma.

EVIDENCES OF DIFFERENTIATION. VARIATION WITHIN A SINGLE ROCK BODY—The physico-chemical facts which lead up to the idea of differentiation have been detailed in Chapter IV; and in the preceding chapter the geological evidence, based on the distribution of igneous rock types in space and time, has been adduced. Other evidence is afforded by variation within a single rock body. Some igneous masses are singularly uniform in composition and texture, most lavas, for instance, and many granite masses. The felsite laccolith of Tinto Hill (Lanarkshire) is strikingly uniform through a thickness estimated at 3500 feet ; [1] and the riebeckite-orthophyre sill of the Holy Isle (Arran) varies not at all through a thickness of 800 feet. In other masses, however, there is often a significant distribution of contrasted parts with regard to the margins. We are here concerned only with variations which have clearly arisen within an originally uniform body of magma, and not with the heterogeneity produced within an igneous mass by successive intrusion of different magmas, by original non-uniformity of the magma, or by the assimilation of foreign rock material.

The contrasted parts within a single rock body may be arranged symmetrically or asymmetrically with regard to the margins of the mass. In the former case sills and dykes will exhibit bilateral symmetry, and laccoliths, stocks, and bosses may show concentric zones of differing composition. The contrasted parts may be sharply separated, but usually a gradual transition can be traced. Most commonly the marginal facies are of basic or mafic composition, whilst the interior parts are more acid or felsic. One of the clearest examples of this mode of arrangement is the gabbro of Carrock Fell (Cumberland).[2] The central part of this mass is a quartz-gabbro of specific gravity 2·85 and SiO_2 averaging 55 per cent. Towards the exterior this passes into an ordinary gabbro with only a little accessory quartz, specific gravity about 2·90, and SiO_2 percentage 48. Finally, at the margin an ultrabasic rock rich in iron-ore is found, in which the specific gravity is 3·26, and the SiO_2 percentage 32·5 (Fig. 47). This arrangement is due to regular variation in the relative proportions of the mineral constituents, labra-dorite, augite, iron-ores, and quartz, the last-named being segregated toward the interior of the mass, and the iron-ores toward the margins.

Cases of symmetrical variation in which a relatively acid or felsic rock is found toward the margin of an igneous mass, while a more basic or mafic rock occupies the interior, are known, but are comparatively rare.

An example of asymmetrical arrangement is given by the

[1] *Summ. Prog. Geol. Surv. Gt. Brit. for* 1924, 1925, p. 102.
[2] A. Harker, *Quart. Journ. Geol. Soc.*, 50, 1894, pp. 311-36.

central part of the Lugar sill (Ayrshire). This is composed of ultra-basic analcite-bearing rocks in which an increasing abundance of olivine from top to bottom of the mass can be traced. The uppermost layer consists of theralite with 13·6 per cent. of olivine. This passes downward into picrite with 30·1 per cent. of olivine, and finally to peridotite at the base with 65·2 per cent. of olivine (Fig. 48).

THEORIES OF DIFFERENTIATION[1]—There are two stages in the process of differentiation, of which the first is the preparation of units such as crystals, liquid sub-magmas, or non-consolute drops of magmatic liquid, by the physico-

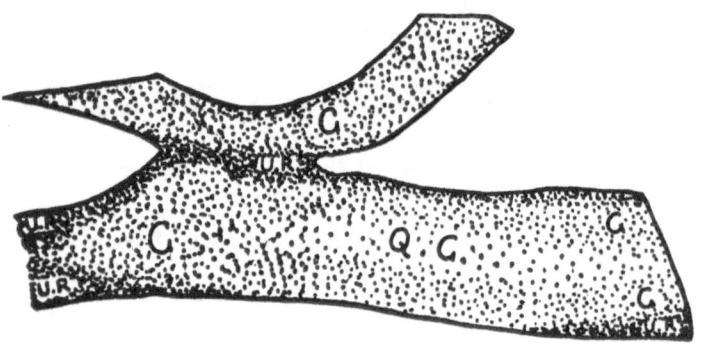

FIG. 47.—MAP ILLUSTRATING DIFFERENTIATION IN THE CARROCK FELL GABBRO.

Scale, 1¼ inches to the mile. Q.G., quartz-gabbro; G, gabbro; U.R., ultrabasic rock. From A. Harker, *Natural History of Igneous Rocks*, 1909.

chemical processes described in Chapter IV. Then follows a stage in which the prepared units are separated more or less completely from one another, and segregated in different regions of the magma chamber, or as distinct masses. This is the geological stage of differentiation, and it is the main subject of the present chapter.

Some petrologists have ascribed differentiation to processes which have taken place in liquid magma prior to the beginnings of crystallisation. Theoretically, the heavier molecules in a mixed solution should segregate in the basal parts of the solution, and the lighter molecules in the upper layers.

[1] For differentiation in general see N. L. Bowen, *The Evolution of the Igneous Rocks*, 1928.

Magmas thus stratified under the influence of gravity played a considerable part in the older speculations regarding differentiation. It is also theoretically possible for a composition gradient to be established by differences of temperature

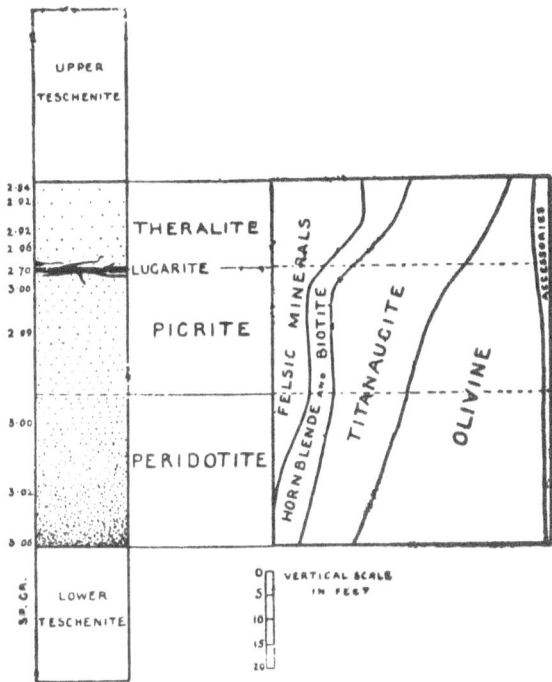

FIG. 48.—VERTICAL SECTION ILLUSTRATING DIFFERENTIATION OF LUGAR SILL.

The increasing density of the stippling represents increasing abundance of olivine crystals downwards. On the right is a diagram illustrating the increase in the amount of olivine in the differentiated central part of the sill. (See text, p. 150.)

in different parts of a liquid magma (Soret effect). Harker, however, has shown that the amount of differentiation produced in this way is practically negligible ; [1] and Bowen

[1] *Natural History of Igneous Rocks*, 1909, p. 316.

remarks that the effects producible by density stratification are probably of the same order of magnitude. At any rate, under ordinary circumstances, crystallisation no doubt super-venes long before any appreciable concentrations are effected, and with crystallisation factors enter, the differentiation effects of which completely mask those due to molecular diffusion.[1]

DIFFERENTIATION BY LIQUID IMMISCIBILITY—It has been thought that differentiation may possibly take place in the magmatic state by the separation of liquid phases of dif-ferent composition (unmixing or immiscibility), just as a mixture of aniline and water, which is perfectly homogeneous above 166° C., separates into non-consolute fractions of aniline admixed with some water, and of water with some aniline, below that temperature. The phenomenon of liquid immiscibility has been invoked to explain the relations of sulphides in silicate magma (Vogt), and of alumina (corundum) in peridotite magma.[2] It has also been used in a general way to explain discontinuous variation in igneous masses, and the juxtaposition of strongly contrasted parts. It is, however, the almost unanimous opinion of petrologists that limited miscibility does not occur between fluid rock-forming sili-cates. Only one instance of this phenomenon has been observed in the thousands of experiments with molten sili-cate mixtures under a great variety of conditions at the Geophysical Institute in Washington (see next page). It has not been found in metallurgical practice, or in other technical operations requiring the use of molten silicates.

Bowen has pointed out that non-consolute fractions separate out first as globules which grow slowly by diffusion, and only collect as separate layers if sufficient time elapses before crystallisation. Were the liquid quenched to a glass, globules of composition different to that of the main mass would be preserved ; and on crystallisation of the hetero-geneous liquid the product should have a very noticeable patchy and blotchy character. If liquid immiscibility occurs in silicate magmas evidence of the above nature should be frequently forthcoming, especially from glassy or partially

[1] " Later Stages of the Evolution of the Igneous Rocks," *Journ. Geol.*, 23, 1915, Suppl., pp. 3-7.
[2] A. Harker, *Natural History of Igneous Rocks*, 1909, pp. 199-200.

glassy rocks.[1] Unequivocal evidence of this nature, however, has not been obtained.[2]

In experimental work at the Geophysical Institute J. W.
Grieg found that silica-rich mixtures of MgO, CaO, FeO,
MnO, etc., with silica, melted to two immiscible liquids.[3]
The temperature of equilibrium between cristobalite and
two liquids was near 1700°, and one of the liquids was
nearly pure silica. Neither the composition of the liquids,
nor the temperature of the phenomenon, were within the
range of even extreme types of igneous rocks.

The interesting suggestion has been put forward that the
presence of abundant water and other volatiles in a magma
might lead to the separation of non-consolute fractions
through supposed limited miscibility between the silicates
on the one hand and the volatile constituents on the other.[3]
The presence of water and other gases has undoubtedly a
considerable effect on differentiation (p. 159), although,
perhaps, not by promoting liquid immiscibility.[4]

The only important case of limited miscibility in rock
magmas is that between sulphides and silicates, which is
illustrated by the limited mutual solubility of slag and sulphide matte in smelting operations. It is possible that some
occurrences of sulphides of iron, nickel, and copper, are due
to the separation of the metallic sulphides in the liquid phase
from magmatic solution, and their subsequent concentration,
still in the liquid phase, by gravitational or flotational processes (p. 161).[5]

DIFFERENTIATION BY CRYSTALLISATION—As soon as a
magma proceeds to crystallisation, the possibility of extensive differentiation is immediately introduced. Differentiation may be brought about by at least two distinct processes :
the localisation of crystallisation aided by diffusion and convection ; and the localised accumulation of crystals in several

[1] N. L. Bowen, *Journ. Geol.*, 23, 1915, Supplement, p. 8 ; *Journ.
Geol.*, 27, 1919, pp. 399-405.
[2] N. L. Bowen, *Journ. Geol.*, 34, 1926, pp. 71-3 ; J. W Greig, *Amer.
Journ. Sci.*, xv, 1928, pp. 375-402.
[3] *Amer. Journ. Sci.*, xiii, 1927, pp. 1-44, 133-54.
[4] J. W. Evans, *Congrès. Geol. Internat., C.R. XXII Session*, Canada,
1913, p. 248 ; Arrhenius, *Geol. Fören. Förh.*, 22, 1900, 395-419.
[5] See Bowen, *op. cit. supra*, p. 422.
[6] By flotation is meant the buoying-up and transference of immiscible
globules of molten sulphides by occluded gases (chiefly H_2S). See
W. H. Goodchild, " Evolution of Ore Deposits from Igneous Magmas,"
Mining Mag., 1918. Reprint, p. 7

different ways, with the concomitant segregation of the liquid magmatic residuum.

Crystallisation may be localised at a cooling margin, where the temperature may be sufficiently lowered for the earliest minerals to appear, while the more central parts of the magma are still liquid. Such crystallisation impoverishes the immediately-surrounding magma in the crystallising substance, but the concentration is supposed to be maintained by free diffusion of that substance from all parts of the magma (Harker), or by convection currents (Becker), with a con·comitant movement of other substances in the opposite direction. By this action the minerals of early crystallisation may be concentrated towards the margins of an igneous mass, producing basic (mafic) selvages around a more acid (felsic) central part consisting of the minerals of late crystallisation (p. 75), and thus bringing about differentiation.

Bowen [1] criticises this conception on the ground that it postulates a much greater freedom of molecular diffusion in silicate magmas than is known to occur. He has shown by experiment that the rate of diffusion is extraordinarily slow. After 256 years, for example, a layer only about 3 inches thick of the first-crystallised minerals would be formed on the margins of an igneous mass, and all the material would have been taken from a marginal band less than 7 yards wide. [2] Furthermore, it is difficult to imagine how the growing crystals (usually heavier than the magma) could remain in position to receive further accessions of substance ; and diffusion to the extent postulated should only result in the inward growth of a few large crystals attached to the walls of an igneous body. Convection currents would lead to a like result, and any loose crystals would be carried away and distributed throughout the magma.

Basic borders may be best explained as the rapidly-chilled marginal parts of a magma, which preserve its original composition ; while the central portions represent the lighter, more felsic differentiate formed by one of the processes of differentiation described below (Fig. 49). In bodies such as sills and laccoliths the chilled borders may completely enclose

[1] " Later Stages in the Evolution of the Igneous Rocks," *Journ. Geol.*, 23, 1915, Suppl., pp. 11-13.
[2] " Diffusion in Silicate Melts," *Journ. Geol.*, 29, 1921, pp. 295-317.

an interior mass in which differentiation has produced several types of rock (Fig. 50).

In Chapter IV it has been shown that there is a progres-

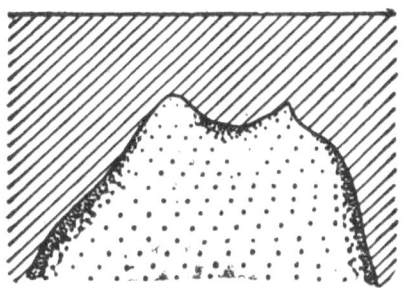

FIG. 49.—DIFFERENTIATION IN STOCKS AND BATHOLITHS.

The diagram shows a stock or batholith with a relatively basic marginal facies (indicated by the increased density of stippling).   After Daly, *Igneous Rocks and Their Origin*, 1914.

sive change in the composition of the minerals formed from a magma as crystallisation proceeds.   The table on page 75 shows the contrasted composition of early-formed crystals in a

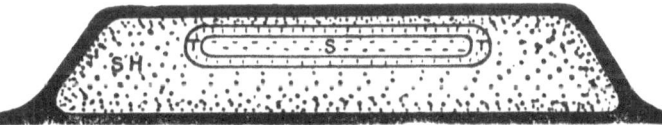

FIG. 50.—DIFFERENTIATION IN THE SHONKIN SAG LACCOLITH.

The diagram illustrates the arrangement of rock types within the lacco-lith.   The black border represents the contact rock (leucite-basalt-porphyry); SH, shonkinite; T, transitional rock; S, syenite.   All the varieties pass gradually into one another.   Vertical scale six times the horizontal.   After A. Harker, *Natural History of Igneous Rocks*, 1909.

common type of basaltic magma, and the residual liquid from which the later minerals are derived.   With maintenance of contact between the early crystals and residual liquid, the contrast of composition is lessened, and the possibility of differentiation diminished; but if, by any means, a separation is effected between early crystals and residual liquid, the difference in composition is intensified owing to the lack of mutual reaction, and separate bodies of rock are produced which are strongly contrasted in composition. In this way arises the well-known parallelism between the course of crystallisation and the course of differentiation. The products of early crystallisation are concentrated at one end of a differentiation series, and the products of later crystallisation at the other end.

Two ways have been suggested by which crystals and residual liquid can be separated: gravitational sinking of the earlier crystals; and straining-off, squeezing-out, or filtration of the liquid by earth pressure.

GRAVITATIONAL DIFFERENTIATION—The sinking of crystals in lavas has been noted by several observers, and Darwin maintained that this process was a prime cause of differentiation.[1] Lane has demonstrated both the sinking and rising of crystals in an extrusive Triassic basalt from Nova Scotia, wherein the felsic minerals tend to be concentrated towards the top, and the mafic minerals towards the base, of the flow.[2] Numerous examples of the subsidence of heavy crystals in intrusive masses have now been described (cf. Lugar sill, p. 150, Fig. 48). Bowen has shown that olivine crystals in silicate melts collect on the bottom of the containing crucible. Similar results were obtained for pyroxenes, but tridymite, when formed, tended to rise to the top.[3]

The minerals mostly involved in gravitational sinking are olivine, pyroxenes, calcic plagioclase, and iron ores, in which the specific gravity at high temperatures is greater than that of the enveloping magma. Of these, olivine seems to be the most important, and the majority of cases in which stratification by density has been described, are due to variations in the proportions of olivine. The Lugar sill (p. 150)[4] is a case in point; and another excellent example is the quartz-dolerite sill of the Palisades of the Hudson River, in which a concentration of olivine has taken place towards, but not at,

[1] *Geol. Observ. Volc. Islands*, 1844, p. 118.
[2] *Trans. Amer. Inst. Min. Eng.*, 1916, p. 535.
[3] *Amer. Journ. Sci.*, 39, 1915, pp. 175-91.
[4] Tyrrell, *Quart. Journ. Geol. Soc.*, 72, 1917, p. 125.

the lower contact, giving rise to a stratum of olivine-dolerite.[1] The rate of subsidence depends on the viscosity of the magma, and on the sizes and shapes of the crystals. Subsiding crystals are not able to sink through the viscous chilled marginal layers to the actual floor of the sill or laccolith, but form a layer somewhat above the base. The size of the crystals is an important factor. Bowen calculates that in granite magmas magnetite crystals 0·1 mm. in diameter would sink no faster than felspar crystals 0·4 mm. in diameter. This fact, together with the brief, and often late, period of crystallisation, and the small quantity of material available, explains why ore minerals, which are much heavier than olivine or pyroxene, are rarely found accumulated towards the base of an igneous mass.

Monomineralic rocks, such as dunite, pyroxenite, and, perhaps, anorthosite, may be formed by this process under especially favourable circumstances. In most cases, however, the minerals of early crystallisation probably sink as a swarm with little tendency to relative movement between the different kinds of minerals. From the common type of basalt magma olivine and magnesian augite may sink together at an early stage of differentiation, giving rise to a peridotite stratum. At a later stage calcic plagioclase may be added, producing a mass of the composition of gabbro. At the same time the segregated residual liquid would, under certain conditions, crystallise into a mixture of alkalic felspars, diopsidic pyroxene (or hornblende and biotite) and quartz, and would thus form a granitic type. If solidification owing to rapid cooling supervenes before the process of gravitative differentiation is completed, intermediate types such as diorite or granodiorite will be formed.

Vogt [2] and many other workers have given reasons for believing that in magma chambers of large size, sunken crystals would be re-dissolved in depth, and would thus form liquid layers of composition not notably different from that of various mixtures of early-formed crystals. On this view the

---

[1] J. V. Lewis, *Ann. Rept. Geol. Surv., New Jersey*, 1907, pp. 125; 129-33.
[2] H. L. Vogt, "The Physical Chemistry of the Magmatic Differentiation of Igneous Rocks," *Vidensk.-selsk. Skr.* 1, *Math.-Nat. Kl. Kristiania*, 1924.

monomineralic rocks, i.e. the rocks enriched in early-formed constituents,[1] are always of very deep-seated origin, having been derived from magmas at very high temperatures, and relatively free from volatile constituents. When intruded into cold rocks they would become completely crystalline in a short distance and long before reaching the surface, which would explain their rarity as dykes, and their non-appearance as lavas.

In the residual liquids are stored up all the mix-crystal components of relatively-low melting-points (p. 75), with excess silica, water, and other volatile constituents, and volatile compounds of these substances with metals, etc. Thus quartz, potassic and sodic felspars, nepheline, leucite, analcite, aegirine, diopsidic pyroxene, muscovite, biotite, hornblende, and numerous rarer minerals and metallic compounds, tend to crystallise from residua of appropriate composition. According to Vogt residual magmatic liquids approximate to eutectic composition (*eutectic-enriched*, or *anchi-eutectic*). He cites acid granites as examples of well-nigh eutectic quartz–alkali-felspar rocks, alkali-syenites as well-nigh eutectic orthoclase-albite rocks, and certain gabbros and norites as well-nigh eutectic plagioclase–pyroxene rocks. In the latter case the term eutectic can only be retained by an extension of its meaning to cover the relations obtaining along equilibrium boundaries between a mix-crystal series (plagioclase), and another independent mix-crystal series (pyroxenes), or an independent mineral such as diopside (p. 70).

FILTRATION DIFFERENTIATION—The subsidence of crystals tends to take place during the earlier stages of crystallisation, when heavy minerals are being formed in a comparatively thin fluid. As crystallisation continues a loose mesh or framework of crystals with residual liquid in the interstices will ultimately be formed. If deformation of the mass occurs at this stage, either by the simple downward pressure of the lifted strata (in the case of a sill or small laccolith), or by the oncoming of a lateral earth pressure, the crystal mesh will be progressively crushed down, and the interstitial liquid will be squeezed out. The liquid will tend to move towards the regions of least pressure, and it may form

---

[1] Vogt has recently termed these rocks *proto-enriched*. *Op. cit. supra.*

bands and schlieren in the portion of the mesh least affected by the pressure, or may become a separate intrusive body.[1] It may be injected as veins or dykes into the already-solidified and cracked marginal parts of the igneous mass. Thus may be explained the red felsite and granophyre veins, dykes, and schlieren found in the quartz-dolerite sills and dykes of the Midland Valley of Scotland, and the irregular bands and veins of lugarite (a teschenitic type very rich in analcite) in the picrite–teschenite sill of Lugar. Pegmatite and aplite dykes associated with plutonic masses may also be regarded as due to the expulsion of residual liquid, not only into the marginal parts of the igneous body, but also into the surrounding country rocks.

In certain lava-like rocks, such as the augite-andesite and tholeiite dykes of the Kainozoic in Western Scotland, acid magmatic residues have often oozed, or have been squeezed, into vesicles, forming spherical spots of glass or of fine-grained material. A further degree of separation results in the intrusion of composite dykes in which a central band of glassy rock (pitchstone) is flanked by tholeiite of earlier intrusion ; or in quite separate intrusions of these rocks.

THE RÔLE OF VOLATILE CONSTITUENTS IN DIFFERENTIA-TION [2]—The nature of the volatile constituents, and their influence on the physical properties of magmas, have been described in Chapter III (p. 51). The importance of the " mineralisers " in igneous rock formation was early recognised, especially by French petrologists. Volatile constituents have a considerable influence on crystallisation, and therefore on differentiation, as they render a magmatic system extraordinarily sensitive to variations in external conditions. With release of pressure the magmatic gases tend to migrate towards points of least pressure and there distil off. If they have entered into combination with certain magmatic constituents to form extremely mobile, if not volatile, compounds, the sort of convection which is caused by the movement of the volatile constituents may be

[1] A. Harker, *Natural History of Igneous Rocks*, 1909, p. 323 ; N. L. Bowen, " Crystallisation-Differentiation in Magmas," *Journ. Geol.*, 27, 1919, p. 406 ; " Differentiation by Deformation," *Proc. Nat. Acad. Sci.*, 6, 1920, pp. 159-62.
[2] S. J. Shand, *Eruptive Rocks*, 1927, Chap. III.

very effective in the transfer of magmatic material.[1]   It has been pointed out by Butler that the valuable metalliferous ore deposits of Utah have been deposited in and around batholiths and stocks which are truncated near their apices by the present erosion surface, while igneous masses planed down to deeper levels have few valuable ore-deposits associated with them (Fig. 51).   This fact is explained as due to the concentration of the mobile volatile constituents carry-

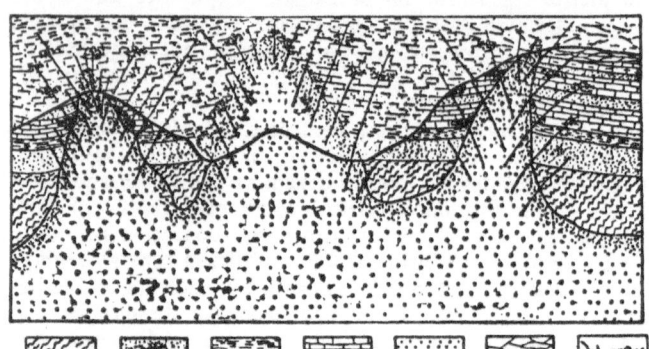

SCHIST   QUARTZITE   SHALE   LIMESTONE   BATHOLITH WITH STOCKS   EXTRUSIVE ROCKS   VEINS AND ORE DEPOSITS

FIG. 51.—DIAGRAM ILLUSTRATING THE EFFECTS OF GAS CONCENTRATION IN BATHOLITHS.

The diagram shows a batholith with three stocks or cupolas rising from its roof, intrusive into a series of quartzites, shales, and limestones. The thick line represents the present surface of the ground. Ore deposits are found at right and left where the surface truncates the apices of stocks. In the centre a basally-truncated stock shows a few veins, but no ore deposits. (See text, p. 160.)   After Butler.

ing metallic sulphides, near the apices of the igneous bodies, by upward migration in the direction of lessened pressure.[2]

Water is unquestionably the most important volatile constituent of magmas. Free hydrogen, oxygen, and sulphuretted hydrogen may also occur. According to W. H.

---

[1] P. Niggli, *Die Leichtflüchtigen Bestandtheile in Magma*, Leipzig, 1920, 272 pp.; " Die Gesteinsassoziationen und ihre Entstehung," *Verh. d. Schweiz. Naturf. Gesell. Neuenberg*, 1920, pp. 1-25; C. N. Fenner, "The Katmai Magmatic Province," *Journ. Geol.*, xxxiv, 1926, pp. 673-772.

[2] *Econ. Geol.*, 10, 1915, pp. 101-22.

Goodchild, the well-known reversible and practically iso-thermal reaction of water with ferrous oxide, resulting in the production of ferric oxide and hydrogen, affords a deli-cately-balanced contrivance for varying the concentration of hydrogen in magmas in response to changing conditions of equilibrium:—

$$2 \; FeO + H_2O \rightleftarrows Fe_2O_3 + H_2.$$

When sulphur or metallic sulphides are also present, they react with water to form sulphuretted hydrogen. The effect of hydrogen, water, and other gases in magmas is to lower the freezing range very considerably, and also to reduce the viscosity. Non-consolute sulphide globules, if present, will be highly charged with magmatic gases, especially H_2S, and may experience a transfer to the cooling margins by a process akin to the metallurgical process of flotation. They may be so buoyed up by the occluded gases as to travel in directions opposed to gravitative descent, although the concomitant downward movement of the early silicate crystals may carry them mainly downwards. Thus Goodchild would explain the concentration of pyrrhotine, pentlandite, and chalco-pyrite in such igneous bodies as the " nickel eruptive " of Sudbury, Canada.[1]

In silicate magmas the pressure which is required to retain any large proportion of volatile constituents in solution must be very great, and even a small amount of crystallisation will result in a considerable increase of vapour pressure. Conse-quently as crystallisation proceeds a " bursting pressure " is produced, which may break through a weak cover, causing vulcanism ; or it may drive residual fluids into the adjacent rocks.[2] This is probably the dominant cause for the extru-tion of pegmatites into the country rocks surrounding a granite mass, although tectonic pressure may aid, or may act independently.

[1] " The Evolution of Ore Deposits from Igneous Magmas," *Mining Mag.*, 1918. Reprint, p. 7.
[2] G. W. Morey, " The Development of Pressure in Magmas as a Result of Crystallisation," *Journ. Wash. Acad. Sci.*, 12, 1922, pp. 219-30. See also W. H. Goodchild, *op. cit. supra*, p. 20, and P. Niggli, *op. cit. supra*, p. 117.

11

PEGMATITE AND APLITE [1]—*Pegmatites* are crystal growths of variable grain size, frequently very coarse, and often with marked intergrowth structures, which are disposed in sheets, veins, and dykes, mainly outside, but also within, a plutonic mass, to which their mineral composition has a general correspondence. Granites and syenites most often have satellites of this character. Pegmatitic growths are found much more rarely in other plutonic rocks, and then generally occur within the igneous body as streaky or pocketty segregations. The minerals of pegmatites are those of the residual stage of the plutonic rocks with which they are associated, and they frequently carry minerals rich in volatile constituents or their compounds. Thus granite pegmatites are mainly composed of alkali-felspars and quartz, but may also be rich in muscovite and hydromicas, and may contain minerals such as tourmaline, topaz, beryl, fluorspar, apatite, and lithia-mica, which are rich in the rarer volatile magmatic substances. Compounds of tin, tungsten, molybdenum, copper, arsenic, bismuth, niobium, uranium, and radium, are occasionally found in granite pegmatites. The pegmatites of syenites and nepheline-syenites often contain minerals consisting of compounds of the " rare-earth " metals, such as zirconium, cerium, lanthanum, along with uranium and thorium.

Pegmatite dykes from granites often show a gradual change of mineral character as they are traced outwards from the plutonic margin. Felspar may almost disappear, leaving a residuum of silica, which may crystallise as an igneous quartz vein.

The above characters suggest that pegmatites are the products of the solidification of final magmatic residua which are especially rich in volatile constituents. These mainly aqueous solutions, in which the rarer and more volatile constituents have been concentrated, are injected into the solidified and cracked margins of the plutonic mass, and into the surrounding country rocks by the development of magmatic pressures, or by earth pressures. As the pegmatite minerals are precipitated, the cooling solutions become richer in water and, perhaps, silica, which may finally be deposited

---

[1] A good recent discussion is given by R. H. Rastall, *Geology of the Metalliferous Deposits*, 1923, pp. 35-47.

as quartz. In some cases they may develop as metalliferous veins; or may emerge as hot springs depositing siliceous sinter, with the juvenile water of which more or less ground water may have mingled.

Aplites are fine-grained, equigranular rocks of allotriomorphic texture (p. 85), which are developed as veins and dykes within plutonic masses, and in the country rocks, but not so abundantly as pegmatites are in the latter. They are most common in connection with granites, but may be found in other plutonic types. Their mineral composition corresponds to that of the final stage of the plutonic magma with which they are associated. Thus granite aplites consist of quartz and alkali-felspars, with occasionally small amounts of muscovite, fluorspar, tourmaline, topaz, etc.

The fine and even grain of aplites is suggestive of comparatively small amounts of volatile constituents in the magmatic residua from which they were derived. The aplite magmas would thus be far less mobile, and would crystallise at higher temperatures than the pegmatites. Hence they would not be expelled to such great distances as the highly-mobile, low-temperature pegmatite fluids. Their relative lack of volatile constituents would also explain the sharp contacts of aplites with the surrounding rocks, in contrast to the permeative boundaries of pegmatites. Aplite magmas are, nevertheless, mobile enough to penetrate the rocks in very thin veins and stringers. The association of pegmatites and aplites as products of the final crystallisation of plutonic magmas, and their occasional injection side by side in fissures,[1] confirms the view that diffusion in silicate melts is relatively very slow, since it does not permit of the uniform concentration even of the highly-mobile volatile constituents, either in the magma chamber or at the injection stage. Poverty in volatile constituents is probably the cause of the uniformity exhibited by some large igneous rock bodies (p. 149).

ASSIMILATION AND HYBRID ROCKS [2]—Another factor leading to non-uniformity in igneous rocks is *assimilation*, in

[1] F. F. Grout, *Econ. Geol.*, 18, 1923, p. 259.
[2] Harker, *Natural History of Igneous Rocks*, 1909, Chap. XIV; Dayl, *Igneous Rocks and Their Origin*, 1914, Chap. XI; N. L. Bowen, "The Behaviour of Inclusions in Igneous Magmas," *Journ. Geol.*, 30, 1922, pp. 513-70.

which is included both the incorporation of foreign rock material in igneous magma, and the commingling of two liquid magmas. Inasmuch as these processes involve the re-mixing of rocks, they represent the reverse of the differentiation process ; and if the mixing is complete and uniform, no heterogeneity will result. The products of complete mixing, however, are seldom recognisable as such, as they are mostly due to deep-seated reactions and are thus rarely exposed by erosion. Alternatively, they may be re-differentiated into a complex of contrasted rock types.

Phenomena of partial absorption and digestion of chips and blocks of intruded rocks are commonly met with about the margins of many igneous bodies, and along the internal contacts of composite dykes and sills. The Kainozoic igneous suite of the West of Scotland, which consists mainly of basic and acid rocks with few of intermediate composition, provides some of the best examples of interaction between acid and basic rocks so far described. In Skye, Rum, Mull, and Arran, all transitions between the mere enclosure of basic rocks in acid magmas without alteration, and complete admixture and distribution of the products of reaction, can be studied (Fig. 52). The products of the intermediate stage of digestion when the xenoliths are partly dissolved, are called *hybrid rocks*. Their presence is indicated by corroded xenocrysts and more or less well-defined patches and streaks of partly-digested material. Mutual reactions occur at this stage by which the invading acid magma is basified, and the basic xenoliths are acidified. Towards the end of the process the magmatic matrix and the xenoliths are so closely assimilated to each other that they are scarcely distinguishable (Fig. 52 c).

New and distinctive minerals are produced by the reactions between magma and xenoliths. Thus the digestion of xenoliths of gabbro, dolerite, or basalt, in acid magma, or of quartzose xenoliths in basic magmas, results in the formation of hypersthene from olivine, and of hornblende from pyroxenes, with the concomitant liberation of iron oxides. Shale inclusions in a basic magma are reacted upon with the formation of aluminous minerals, such as corundum (sapphire in Mull), spinel, sillimanite, cordierite, and anorthite.[1] Whole-

[1] H. H. Thomas, *Quart. Journ. Geo. Soc.*, 78, pt. 3, 1922, pp. 29-60.

sale enclosure of Dalradian quartzites and mica-schists within a large mass of norite in the Arnage district of Aberdeenshire

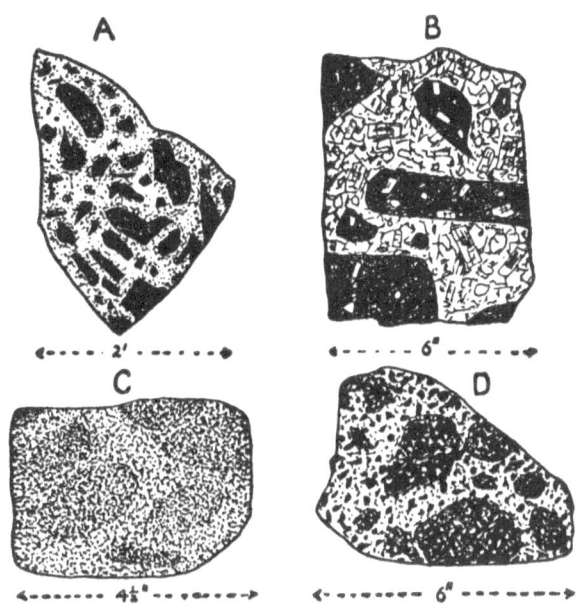

FIG. 52.—HYBRID ROCKS FROM ARRAN.

A. Basalt fragments in matrix of felsite. Summit of Ross Road, Lamlash. Little intermixture.

B Hypersthene-basalt included in quartz-porphyry, Bennan. The basalt carries xenocrysts of quartz and felspar derived from the quartz-porphyry by some degree of admixture before intrusion.

C. Mixture of basalt and more acid type (craignurite), The Sheans, Gleann Dubh. There has been much intermixture; the fragments are well-rounded, and but little differentiated from the matrix.

D. Basic diorite fragments in acid diorite matrix. Head of Gleann Dubh. Detachment and floating-off of crystals.

has led to the formation of a zone of hybridised or *contaminated* rocks, full of xenoliths in all stages of digestion, and with the formation of minerals such as cordierite, spinel, and

garnet, which are unusual or abnormal in igneous rocks.[1]
Very similar cordierite-rich hybrid rocks have been described
from the base of the Bushveld complex near Pretoria, where
norite has assimilated shaly sediments.[2] Read has shown
that the contamination process depends upon reciprocal
reaction between the gabbroid magma and the argillaceous
xenoliths, whereby the magma becomes more acid and the
xenoliths more basic. The xenoliths become richer in lime
and magnesia, and the magma in alumina, potash, and soda.
Furthermore, if such reactions take place towards the top of
a gabbro mass, as the xenoliths become progressively heavier
and the magma lighter, the xenoliths must sink and accumulate near the floor of the magma chamber. In this way a
pronounced differentiation may be effected, resulting in the
production of a magnesia-lime-rich basal layer (basic norite,
pyroxenite), and an alumina-alkali-rich upper part (granitic).[3]

By interchange with the magmatic silicates and the loss of
carbon dioxide, the inclusion of limestone within an igneous
magma results in the formation of lime-rich pyroxenes, amphiboles, and plagioclases, and occasionally lime silicates such
as wollastonite, zoisite, vesuvianite, etc. A large amount of
lime, silica, or alumina, may be incorporated into a magma
without essentially changing its mineralogy; for these substances belong to the same multicomponent system of rock-
forming oxides as is represented by an igneous magma. The
addition of lime to basalt magma tends, by reaction, to
increase the amount of olivine and magnetite, and to make
the pyroxenes and plagioclase somewhat richer in lime than
they otherwise would be.[4] Superheated basalt might, indeed,
be able to form melilite by the solution of calcium carbonate,
thus giving rise to melilite-basalt. The possibilities of reactions between granite and limestone are dealt with below.

The reactions between igneous magmas and other igneous
rocks or sediments are governed by the principle that a liquid
saturated with a certain member of a reaction-series (p. 76)
is effectively supersaturated with all preceding members, and

[1] H. H. Read, *Quart. Journ. Geol. Soc.*, 79, pt. 4, 1923, pp. 446-86.
[2] A. L. Hall and A. L. du Toit, *Trans. Geol. Soc. South Africa*, 26, .923, pp. 69-97.
[3] *Geol. Mag.*, 61, 1924, pp. 433-44.
[4] Bowen, *op. cit supra*, pp 543-50.

unsaturated for all succeeding members. Thus a granite magma which is precipitating biotite is effectively supersaturated with olivine, pyroxene, and amphibole, and therefore cannot dissolve them ; but it can, and does, react with them in such a way as to convert them into biotite, the phase with which the magma at the moment is saturated. The action of basic magma upon acid inclusions which consist of later members of the reaction-series than the one with which the magma is saturated, is a kind of reactive solution. The inclusions are gradually dissolved with the concomitant precipitation of the crystalline phase with which the magma at the moment is saturated. If these crystals are removed by gravity or otherwise, the crystallisation continues on its normal course towards a final acid phase, the amount of which is augmented by the amount of the inclusions. This result is contingent upon slow cooling ; but the solution of the inclusions is not essential to the formation of the granitic differentiate, which would have appeared, although in smaller amount, if the inclusions had not been incorporated.

ORIGIN OF ALKALINE ROCKS [1]—The distinctive mineralogical characters of alkaline rocks have already been pointed out (p. 137). A deficiency in silica (and/or alumina) relatively to alkalies, resulting in the formation of low-silication minerals such as nepheline or leucite (p. 49), is the main chemical feature of the group. It has been pointed out that truly alkaline rocks are remarkably rare in comparison with the calc-alkaline groups, and probably constitute less than 1 per cent. of the whole.[2] On the other hand, they are very widely distributed. The bulk and distribution of alkaline rocks suggest that there are no large reservoirs of alkaline magma, but that they are formed by local differentiation under special conditions from more abundant magma types.

The origin of alkaline rocks in general, and whether or not they have been derived from the overwhelmingly more abundant calc-alkaline magmas, are much-debated questions. In the first place the normal process of crystallisation-differentiation leads to the segregation of the bulk of the

[1] For a general discussion see C. H. Smyth, *Proc. Amer. Phil. Soc.*, 66, 1927, pp. 535-80.
[2] R. A. Daly, *Igneous Rocks and Their Origin*, 1914, pp. 46-52.

alkalies in the residual magma, along with excess silica and the volatile constituents. If the silica can be got rid of by filtration, or by combination with foreign rock material, the concentration of the alkalic compounds may be so augmented as to lead to their precipitation.

At the granitic stage of crystallisation the following equilibria may be assumed to exist in the magmatic liquid, especially in the presence of abundant water and other volatiles :—

$$KAlSi_3O_8 \rightleftharpoons KAlSiO_4 + 2SiO_2$$
$$\text{(orthoclase)} \quad \text{(kaliophilite)} \quad \text{(quartz)}$$
$$NaAlSi_3O_8 \rightleftharpoons NaAlSiO_4 + 2SiO_2$$
$$\text{(albite)} \quad \text{(nepheline)} \quad \text{(quartz)}$$
$$2\{(FeMg)O, SiO_2\} \rightleftharpoons 2(FeMg)O, SiO_2 + SiO_2$$
$$\text{(pyroxene)} \quad \text{(olivine)} \quad \text{(quartz)}$$

The precipitation of orthoclase, albite, pyroxene, quartz, and of biotite (by a combination of kaliophilite, olivine, and albite or nepheline molecules) will then lead to the concentration of $NaAlSiO_4$ (nepheline) and volatile constituents in the residual magma. If, now, in some way the precipitated minerals are removed from contact with the magma, the concentration of $NaAlSiO_4$ may reach the stage at which nepheline is formed. At the same time the concentrations of CO_2, S, SO_3, Cl, etc., may be sufficient to cause the formation of minerals such as cancrinite, nosean, haüyn, and sodalite, which are peculiar to alkaline rocks.[1] The bulk of alkaline rock formed in this way from calc-alkaline magma would be very small, thus matching the observed proportions. Further, the alkalic member of an igneous complex must be the youngest, a deduction which corresponds with the observed facts of age sequence. In the Bushveld complex of the Transvaal, for instance, the order of intrusion is norite, granite, syenite, nepheline-syenite.

Professor R. A. Daly [2] has advanced the now well-known theory that alkaline rocks rich in felspathoids are due to the interaction of calc-alkali magma (mostly basalt) with calcareous sediments. The binding of a considerable amount

[1] N. L. Bowen, "Later Stages of the Evolution of the Igneous Rocks," *Journ. Geol.*, 23, 1915, Suppl., pp 44-6; 55-61.
[2] *Bull. Geol. Soc Amer.*, 21, 1910, pp. 87-118; *Igneous Rocks and Their Origin*, 1914, Chap. XX; *Journ. Geol.*, 26, 1918, pp. 97-114

of silica by the absorbed lime, and the removal of the products by differentiation processes, is thought to lead to such
desilication of the remaining magma as to cause the formation of felspathoids instead of felspars. The introduction of
carbon dioxide at the same time would facilitate the separation of the alkalic residual liquor, and might explain the origin
of cancrinite and " primary " calcite in certain alkalic rocks ;
although the occurrence of sulphur-bearing minerals as haüyn
and nosean, and of chlorine-bearing minerals as sodalite is
left unexplained.

This view has been criticised on the grounds that the
common association of alkaline igneous rocks with limestones,
as cited by Daly, is purely fortuitous ; and that reactions
between limy sediments and saturated basalt are endothermic,
i.e. heat is absorbed, and the amount of heat necessary for
the transformation of carbonate into silicate could only be
supplied by the crystallisation of the phase with which the
magma was already saturated, olivine or pyroxene. The
net results of the absorption of lime are then the precipitation
of olivine, and of pyroxenes and plagioclases richer in lime
than they otherwise would be, with the hastening of final
solidification.[1] Even a considerable amount of superheat in
the original magma would be used up in this way.

The function of volatile constituents as one of the most
important factors in the segregation of alkaline fractions from
calc-alkaline magmas has been emphasised by several writers,
notably C. H. Smyth,[2] and S. J. Shand.[3] They point to the
frequent and abundant occurrence of minerals containing
H_2O, CO_2, S, Cl, F, and P, in alkaline rocks. Shand notes
that richly alkaline rocks often occur in cupolas, volcanic
pipes, and fissures, in which a long-continued streaming of
magmatic gases might be expected to have taken place.

Shand's view is that crystallisation-differentiation may
produce a magmatic residuum, at once siliceous and alkaline,
which may crystallise to form an alkali granite. If this
magma, however, is developed in contact with limestone, it

---

[1] N. L. Bowen, "The Behaviour of Inclusions in Igneous Magmas,"
Journ. Geol., 30, 1922, p. 546. See also Bowen, *Journ. Geol.*, 23, 1915,
Suppl., p. 61 ; *ibid.*, 27, 1919, pp. 394-9 ; 426-30.
[2] *Amer. Journ. Sci.* (4), 36, 1913, p. 46.
[3] *Proc. Geol. Soc. South Africa*, 1922, pp. 10-32.

may be so desilicated by the consequent reactions as to produce highly-alkaline rocks such as nepheline-syenite, or even ijolite and urtite.   Re-distribution of the components would be facilitated by the presence of the abundant volatile constituents which are always concentrated towards the close of a magmatic episode, and which would be reinforced by carbon dioxide from the absorbed limestone.   The addition of lime to the freezing magma would then explain the abnormally late crystallisation of certain lime-rich minerals such as melanite, schorlomite, sphene, apatite, wollastonite, pectolite, vesuvianite, cancrinite, and calcite, in alkaline rocks.

PART II

THE SECONDARY ROCKS

CHAPTER IX

INTRODUCTION

GENERAL—The rocks included in this division are those which have been formed by the chemical or mechanical activity of the agents of denudation on pre-existing rocks, and which have been deposited at ordinary temperatures and pressures from suspension or solution in water or air. They are the products of the secular decay and disintegration of the rocks of the earth's crust. The broken debris, and most of the dissolved matter arising from these processes, are transported by wind and water, and deposited in the hollows of the land surface and of the sea floor. Some of the material is left as a residual mantle within the area in which the reactions take place ; and while the land waste as a whole may be conceived as on the march to its ultimate resting-place, the sea, much of it remains long enough on the land to build up massive formations several thousands of feet thick. The material carried in suspension is deposited when the velocity of the transporting medium is checked or its physical condition otherwise changed ; the dissolved materials are either precipitated directly by some change in the physical or chemical conditions of the media, or indirectly by the vital activities of animals and plants. The secondary rocks have been accumulated under a great variety of conditions, and consequently show great variations in mineral and chemical composition and in texture.

THE BREAKING-DOWN OF THE ROCKS[1]—Under the influence of the various agents of denudation the minerals and rocks of

[1] J. W. Evans, "The Wearing Down of the Rocks," *Proc. Geol. Assoc.*, xxiv, 1913, pp. 241-300 ; xxv, 1914, pp. 229-70.

the earth's crust tend to break up into finer and finer particles, and also partly to go into solution. The breaking-
down is accomplished by the processes of *decomposition* and
disintegration. In decomposition the minerals of the rocks
are acted upon by air and water; chemical changes take
place; the soluble products are carried away by water; and
the chemically-simpler and more durable residue is left in
place. In disintegration the rock is broken up without
chemical change by the disruptive effects of changes of temperature, frost, abrasion by ice, water, or air carrying sand.
The result of both processes of decay is that the rocks are
broken down into finer and generally more durable material,
and that some portion enters into solution. The first product of these changes is a mantle of broken and decomposed
material of varying composition and thickness, called the
regolith (mantle rock), which covers the whole surface of the
earth except in areas in which it is removed as fast as it is
formed. The regolith may remain in place for a long period,
or it may at once be attacked by transporting agencies, to
find, after few or many halts, its ultimate resting-place in
the sea.

Disintegration and decomposition usually occur together,
but one process is generally dominant. Decomposition is
more active in moist, warm, low-lying areas; and disintegration occurs mainly in the drier, higher, and colder regions
of the earth's surface. The sum total of the results of decomposition and disintegration is known as *weathering*.

DECOMPOSITION OF ROCKS—The principal agents of decomposition are water and air. When rain falls through the
atmosphere it dissolves a certain proportion of the carbon
dioxide, oxygen, and other gases. This oxygenated and
carbonated water is especially active in attacking the minerals. It is reinforced by the ground-water, which has
already attacked the rocks, and is therefore poorer in oxygen
and carbon dioxide, but richer in dissolved substances which
may exert a very active influence in further attack upon the
rock constituents. It may be especially rich in acid sulphates
derived from the solution of pyrites, in organic acids from
vegetable decay, and in alkaline carbonates, all of which
increase its chemical potency.

The chief processes of decomposition are solution, oxidation, hydration, and carbonation. Nearly all minerals are reacted upon to some extent by water, especially when it contains the above-mentioned substances. Some, however, are much more susceptible than others ; and minerals may thus be divided into those which are relatively resistant, such as quartz, muscovite, and zircon ; and those which are altered with comparative ease, such as the felspars and most of the ferro-magnesian minerals. The process of oxidation involves the alteration of minerals with the production of oxides. It is especially active with iron-bearing minerals, forming the iron oxides hæmatite and limonite, which are the chief colouring matters of rocks and produce the red, brown, and yellow tints which are so common on weathered surfaces. Hydration is a process by which minerals are altered into substances rich in combined water. Magnesium-bearing minerals, such as olivine, are thus altered into serpentine and talc ; biotite and other ferro-magnesian minerals are broken down into chlorite, and felspars are decomposed with the formation of hydrous aluminium silicates and free silica. In carbonation the minerals are altered with the formation of carbonates. Many minerals are liable to this mode of decomposition, but it is especially effective with those containing the alkali metals sodium and potassium, as well as calcium and magnesium. Subsidiary modes of alteration may produce various sulphates and chlorides.

The effect of decomposition, therefore, is to produce certain soluble substances such as carbonates, sulphates, and chlorides, and to leave behind an insoluble residue consisting of hydrated oxides and silicates, mixed with minerals such as quartz and white mica, which have suffered little or no attack by the agents of weathering.

As a concrete example we may describe in detail the decomposition of an ordinary granite containing quartz and orthoclase, with a subordinate quantity of oligoclase, muscovite, and biotite, and zircon and apatite as accessory minerals.

Quartz	SiO_2	Remains undecomposed	Sand grains.
Orthoclase	K_2O	Goes into solution as carbonate, chloride, etc.	Soluble material.
	Al_2O_3 $6SiO_2$	Hydrated and combined to form hydrous Al. silicate, with the liberation of soluble silica	Clay.
			Soluble material.

Oligoclase	$3Na_2O$	Goes into solution as carbonate, chloride, etc.	Soluble material.
	CaO	Forms carbonate, which is soluble in water containing carbon dioxide	Soluble material.
	$4Al_2O_3$ $20SiO_2$	Decomposes as in orthoclase	Clay.
Muscovite	$2H_2O$ K_2O $3Al_2O_3$ $6SiO_2$	Remains undecomposed	Mica flakes.
	H_2O	. . . . .	Water.
	K_2O	Goes into solution as carbonate or chloride . . .	Soluble material.
Biotite	$2(Mg, Fe)O$	Goes into solution as carbonate or chloride; iron carbonate oxidises to hæmatite or limonite .	Soluble material and colouring matter.
	Al_2O_3 $3SiO_2$	Forms hydrous Al. silicate, and soluble silica . .	Clay. Soluble material.
Zircon, ZrO_2, SiO_2		Remains unaltered . .	Zircon grains and crystals.
Apatite, $Ca_3(PO_4)_2$, (F, Cl)		Is soluble	Soluble material.

Hence the decomposition of a granite furnishes many different kinds of material, which may be listed as follows :—

(a) *Unaltered Minerals*, including quartz and zircon, which form *sand grains ;* and muscovite, which produces *mica flakes*.

(b) *Insoluble Residues*, including the hydrous aluminium silicates which are the fundamental constituents of *clays ;* and iron oxides, which are the *colouring matters* of rocks.

(c) *Soluble Substances*, including salts of potassium, sodium, calcium, magnesium, iron, etc., and silica.

The soluble material generally finds its way at once into the rivers and is carried to the sea, contributing to the dissolved salts of the ocean. Occasionally, where evaporation can take place, these salts may be deposited at an early stage of their seaward journey. They cannot long remain, however, save in a rainless climate, when they may accumulate to form valuable saline deposits. Soluble silica is usually quickly re-deposited in veins and fissures, and as cementing material in rocks. The insoluble products and the unaltered minerals may remain for a time in place, forming part of the regolith, but ultimately they are carried into the rivers and thence to the sea. The bulkier sand grains are dropped first and form beds of sand ; the finer clay and silt particles are carried farther out to sea to form mud-banks. The mica flakes may settle indifferently in either deposit. All three may be deposited temporarily along the course of a river, or in a lake, forming beds of sand and mud.

The decomposition of basic rocks proceeds on the same general lines as that of granites ; but since they are richer in ferro-magnesian silicates, they produce more soluble material and iron oxides, but less free silica and clayey matter than granites.

DISINTEGRATION OF ROCKS—In high mountains, desert regions, ice-covered and snow-covered areas, the process of decomposition is largely in abeyance, and disintegration is the dominant mode of breaking-down of the rocks. In polar and mountainous regions chemical change in minerals is retarded by the prevalent low temperature. In deserts the solutions, by reason of heat and great concentration, may exert a powerful but localised effect on the rocks ; but as the solutions are generally in small amount, their effects are entirely overshadowed by those due to mechanical disintegration.

Disintegration may result from a variety of causes. The great diurnal variations of temperature in desert and mountainous regions cause strains to be set up in the surface layers of rocks, by which fragments are scaled off (exfoliation). Pebbles may be split into numerous thin, wedge-shaped pieces by this action.[1] The freezing of water in fissures tends to disrupt rocks into angular fragments, and much of the weathering in high mountains takes place in this way, the summits being covered with a thick layer of rock debris. The abrasive action of sand carried by wind or water causes the disintegration of rocks in deserts, or in the channels of swift-running, sand-laden rivers. Glaciers may pluck and rend boulders from their beds ; and by their slow resistless movement grind the material they carry against the sides and floors of the containing valleys, with the formation of sand and mud. Most streams issuing from glaciers are

[1] M. S. Johnson, *Proc. Geol. Assoc.*, xxvi, 1915, p. 150.

heavily laden with material derived from this action. The pounding of waves may result in much disintegrative action, as is proved by extensive coast erosion. Finally, organic agents often have a marked mechanical effect upon rocks. The roots of plants prise open the fissures in rocks in their search for moisture and nourishment; burrowing animals turn over the soil and subsoil; and man himself, by tilling the ground, deforestation, tunnelling, quarrying, mining, and in numerous other ways, helps to disintegrate the rocks.[1]

Disintegration usually occurs under conditions which preclude much chemical activity upon the rocks. Consequently the products of disintegration are frequently quite fresh or comparatively unaltered rock and mineral fragments. On the other hand, disintegration, by breaking up the rock into smaller fragments, helps to expose an enormously greater surface to the agents of decomposition.

By disintegration a granite will break up into a coarse sand composed of fragments of quartz, felspar, and mica, mixed with pieces of rock not yet broken down into the component minerals. Many granite areas carry sands of this composition, which are called *arkose-sands*, or when consolidated, *arkose*. The sand of Sannox Bay, Arran, which is close to the great northern boss of granite in that island, is composed of quartz, fresh felspar, and biotite, obviously derived from the disintegrative weathering of the granite. A basic rock broken up in the same way gives rise to a rock called *wacke* or *graywacke*, which is composed of plagioclase felspar, ferromagnesian silicates, and quartz. The latter mineral is frequently quite abundant in graywackes, and its presence is due to the fact that quartz is by far the most abundant resistant mineral, and finds its way into all kinds of sedimentary deposits.

Disintegration may also produce rough angular rubble consisting of any kind of rock, which may mantle a mountain top, or accumulate by the action of gravity at the foot of a slope. These accumulations are called *talus* or *scree* when unconsolidated, and *breccia* when welded or cemented into a coherent mass.

The resultant of the twin processes of decomposition and

[1] R. L. Sherlock, *Man as a Geological Agent*, 1922, 372 pp.

disintegration is *weathering*, and the product thereof, in the first place, is the mantle of loose, broken, and largely decomposed material, the *regolith*, which covers the surface of the earth. The term regolith was first proposed by Merrill [1] to cover the sum total of terrestrial accumulations, including not only the sedentary or residual products of weathering, but also the material transported by the action of gravity, rivers, glaciers, and wind. Chamberlin and Salisbury, [2] however, restrict the term to " the loose matter that springs from rock decay, wear, and fracture, and other forms of disintegration." The finely-broken upper layer of the regolith, well-aerated, and mixed with decayed organic matter, is the *soil* (p. 184).

TRANSPORT—The soluble or insoluble material supplied by weathering is either accumulated in place, or is transported and deposited elsewhere. The agents of transport are rivers, waves, ocean currents, wind, and glaciers. Rivers may carry material in solution or suspension, or may roll it along their beds. A portion of the soluble products of weathering, after travelling for a longer or shorter period in the ground-water, is finally discharged into the rivers and carried to the sea. The finer insoluble material, sand and clay, likewise ultimately finds its way into the streams, and is deposited, after many halts, in the sea. Coarser material which the current is unable to lift is rolled bodily along the bed.

The waves and currents of the sea also shift the materials supplied by the rivers, or obtained by sea-wear along the coasts. The wind is a most effective agent of transport, as is shown by the universality of dust. A single storm over the central United States on 9 March, 1918, transported over a million tons of dust a thousand miles.[3] The *loess* of China is believed to be simply an extensive deposit of wind-blown dust derived from the Asian deserts. The dunes of deserts and sandy shores are further examples of the power of the wind in carrying material from one place to another. Finally, glaciers carry broken rock material on their surfaces, frozen within the body of the ice, or dragged along beneath it. The

[1] *Rocks, Rock-weathering, and Soils*, 1897, p. 299.
[2] *Geology*, vol. i, 2nd ed, 1909, p. 422.
[3] A. N. Winchell and E. R. Miller, *Amer Journ. Sci.*, 46, 1918, pp. 599-609.

12

great thickness and extent of the boulder-clay and moraines due to the Pleistocene glaciation of the Northern Hemisphere testify to the transporting power of ice.

The general effect of mechanical transport is to break up the transported material into finer particles and to round off the sharp edges of the fragments. Wind is by far the most efficient agent of rounding, and grains which have suffered long transport by wind show almost perfect spherical forms (millet-seed grains). Ice transport, however, permits of very little rounding. The more mobile the agent of transport the more efficiently does it sort the material it carries. The heavier and larger particles are dropped in one place, the finer and lighter ones being carried on greater distances. Thus beds of comparatively pure sand and clay are deposited separately. Wind, again, is the most efficient sorter of the grains, and deposits carried by wind are often characterised by their homogeneity and uniformity. In ice transport, however, there is little or no assortment of the material, and on the melting or retreat of the ice, it is dumped down into an unassorted and heterogeneous mixture of rock-flour, grains, pebbles, and boulders of all sizes. A classification of the material deposited upon the earth's surface may thus be made into (1) Transported; and (2) Sedentary.

DEPOSITION—The ultimate destination of transported material, whether carried by water, wind, or ice, is the sea; but it may be temporarily deposited on the land, and the deposits thus formed may persist for several geological periods before they resume their march to the sea. This leads to a distinction between *continental* and *marine* deposits.

Deposition may be either mechanical or chemical, according to whether it affects the mechanically-transported insoluble material, or the substances carried in solution. The material carried in suspension or in other ways by water, wind, and ice, is deposited when the transporting medium is overloaded, when its velocity is checked, or when it suffers a chemical or physical change. Very extensive deposits of clay, silt, and sand thus occur in the lower parts of river systems, and also where rivers debouch into the sea (deltas). The settlement of material entering the sea is aided not only by the decrease in velocity of the river current, but also by the admixture of salt water, which promotes a physical change (flocculation) favourable to the deposition of suspended material.

The soluble matter derived from weathering may be deposited either on land or in water, directly by physicochemical processes such as precipitation and evaporation, or indirectly by the agency of organisms. If water containing dissolved substances meets a current containing different substances, chemical reactions may take place, resulting in the precipitation of material. Furthermore, evaporation of the solution may occur until it becomes saturated with the dissolved substances, when deposition takes place. A large proportion, however, of dissolved substances resulting from rock decay reaches the ocean, where their accumulation during geological time is held to be the cause of the present salinity of sea water.[1] Some of the salts, especially those of calcium and magnesium, are extracted by living organisms in order to build their shells and skeletons. Upon the death of the organisms their hard parts accumulate and sometimes form extensive deposits. The vital activities of living organisms, especially certain lowly forms of plants, cause the precipitation of material from solution. Thus the activity of bacteria is believed to be effective in the deposition of bog-iron-ore in swamps and lakes, and algæ in the formation of certain calcareous rocks. Interruption of the processes of deposition, or repeated changes in the composition of the materials deposited, which may only be slight, are responsible for the *bedding* or *stratification* which is so characteristic of sedimentary rocks (p. 196).

THE CLASSIFICATION OF THE SECONDARY ROCKS—The Secondary Rocks are due to the distribution of the products of weathering according to the following scheme :—

---

[1] J. Joly, *The Birth-time of the World*, 1915. " Denudation," p. 30.

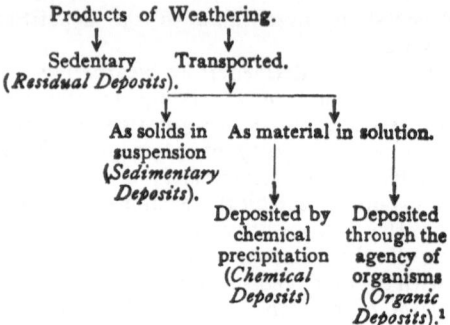

Subdivision into the four groups of Residual, Sedimentary, Chemical, and Organic, therefore seems to afford the most natural and logical mode of classification. These groups, however, must not be thought of as sharply defined; they pass imperceptibly into one another; and as the factors concerned in the deposition of the Secondary Rocks are multifarious, and more than one may be instrumental in the deposition of any given type, it follows that there is often difficulty in assigning some rocks to any natural group. The following chapters deal with each of the above-mentioned classes in order, beginning with the Residual Deposits.

[1] Such materials as plant-accumulations (peat, coal, etc.) can hardly be said to be entirely derived in the manner here suggested, but they are nevertheless most conveniently included under the Organic Deposits (p. 243).

CHAPTER X

THE RESIDUAL DEPOSITS

RESIDUAL DEPOSITS IN GENERAL—The *residual deposits* are the insoluble products of rock weathering which have escaped distribution by transporting agencies, and which still mantle the rocks from which they have been derived. Their components belong to two classes, namely, unaltered minerals from the original rocks, and the insoluble products of decomposition. The nature of the minerals of the first class depends upon that of the bedrock, but quartz, felspars, and muscovite are amongst the commonest, although the felspars are usually much decayed. The rarer durable constituents of rocks, such minerals as zircon, rutile, garnet, tourmaline, kyanite, etc., and various iron oxides, magnetite, hæmatite, ilmenite, and chromite, are also to be found. The silicate minerals comprised in the second class are chiefly hydrous aluminium silicates of the kaolinite-halloysite group, hydrous magnesium silicates of the serpentine-talc group, chlorites and hydromicas, zeolites, and the epidote minerals. Various hydrated oxides of iron and aluminium, and colloidal silica, may also be present.

Since these materials have not been affected by transport they are naturally unsorted and angular. There will be a mixture of fragments of those sizes which are consonant with the nature and grain-size of the weathered rock. Plutonic rocks such as granite will give rise to fragments of sand-grain size, since the fractures generally take place around the individual mineral grains. The fragments will usually be sharply angular for the same reason. The disintegration of conglomerates and breccias will produce coarse gravelly material. On the other hand, the insoluble products of rock decay are usually of flour-like fineness, and may thus, if present, form a matrix for the coarser materials.

It is necessary to state that, in this discussion, the term *insoluble* is used in a relative sense; for the maximum redistribution that can take place in a residual deposit is through partial solution of the "insoluble" materials by further intensive weathering. The products frequently pass into the colloidal state (p. 219), and are re-precipitated almost in place, with the result that oolitic, pisolitic, and other concretionary structures, are common in residual deposits.

TERRA ROSSA; CLAY-WITH-FLINTS—Limestone country in arid regions is often covered with a reddish clayey soil, the *terra rossa*, which is the insoluble residue of clay and other mineral matter left behind after solution of the limestone. The Karst country of the Adriatic, Istria and Dalmatia, provides one of the best-known examples of this phenomenon. In regions of somewhat larger rainfall the terra rossa is washed into depressions, swallow-holes, and caves, soon after formation, leaving bare limestone on the uplands. The bones of Pleistocene animals, and relics of primitive man, are often found buried in this deposit.

The *clay-with-flints* of the South of England is a formation the main constituents of which are probably due to the same cause as terra rossa. It is a reddish clay, mingled with unworn and broken flints, and with a few rounded quartz grains, which mantles the surface of the chalk in many localities. All of these constituents are probably the residues left after the solution of the chalk. Even the rounded quartz grains are probably derived from that formation, as they have been found in it over an area stretching from France to the West of Scotland.[1] In some places the clay-with-flints may have been worked up with the overlying Eocene sands and clays. Sherlock and Noble have suggested that the Pleistocene ice sheet was the agent responsible for the mixing and distribution of this material in the counties north of the Thames.[2]

LATERITE AND BAUXITE[3]—*Laterite* is a reddish, porous, concretionary material which covers vast areas in tropical and sub-tropical lands, forming a hard surface crust on iron-rich and aluminium-rich rocks. It varies greatly in composition, but in general it consists of a mixture of hydrated

[1] E. B. Bailey, "The Desert Shores of the Chalk Sea," *Geol. Mag.*, 1924, p. 102.
[2] *Quart. Journ. Geol. Soc.*, lxviii, 1912, p. 199.
[3] C. S. Fox, *Bauxite*, 1927, 312 pp.; H. Harrassowitz, *Laterit*, 1926, 311 pp.

ferric oxide with hydroxide of aluminium in various proportions, frequently also with manganese dioxide, titanium
dioxide, and free silica. The name is derived from *later*
(Latin, a brick), and is a reference to its employment for
brick manufacture in India, where advantage is taken of its
property of hardening upon exposure. When the aluminous
constituent predominates, the colour lightens to yellowish
or whitish, and the rock becomes earthy and clay-like, occasionally with oolitic and pisolitic structures. This is the rock
which is called *bauxite* (after Beaux, in the South of France).

Laterite and bauxite appear to be due to intensive weathering under extreme oxidising conditions in tropical climates
characterised by strongly-contrasted wet and dry seasons.
Under ordinary conditions of weathering in humid temperate
climates, the silicates of the alkalies, lime, and aluminium
(e.g. the felspars), as we have seen (p. 173), lose their bases
and gain water, forming hydrous aluminium silicates of the
general composition $Al_2O_3,2SiO_2,nH_2O$. The iron present is
mostly converted into ferrous salts which are carried away
in solution. Hence the residues under these conditions are
comparatively poor in iron but are rich in combined silica.
But under the above-stated tropical conditions, decomposition
of the silicates goes a stage further. They lose not only their
bases but all the silica too, so that only the hydroxide of
aluminium is left. Moreover, under the strongly oxidising
conditions, the iron in the rock produces ferric salts, and these
are not only less soluble than ferrous salts, but are very easily
oxidised to ferric oxides and hydrates which are then precipitated.

Obviously laterite will be produced from iron-rich, and
bauxite from aluminium-rich, materials. Some geologists
have suggested that iron solutions are drawn up to the
surface by capillary action during the drought season, with
the result that iron oxide is deposited, and may thus make an
important contribution to laterite formation.[1] This process
is the same as that which produces the hard layer known as
the " iron-pan " below the soil in some sub-tropical regions.
Sir T. H. Holland has suggested that bacteria are concerned
in the fixation of the iron in laterite, but this view has not

[1] W. G. Woolnough, *Geol. Mag.*, 1918, p. 385.

yet been confirmed.[1]   Many observers are agreed that laterisation and bauxitisation can only take place when the rocks
are in contact with ground-water as well as the atmosphere.
J. M. Campbell has shown that laterite ceases to form when
for any reason it passes out of the ground-water zone; and
when laterite is faulted below the level of the vadose water,
the iron is leached from it, resulting in the relative concentration of aluminous material, and the production of bauxite.[2]

Bauxite is the principal ore of aluminium, and iron is often
sufficiently concentrated in laterite to form serviceable iron
ore (e.g. Mayari ores of Cuba).   Nickel, cobalt, manganese,
and gold, are other metals which are sufficiently concentrated
by laterisation from rocks of suitable original composition to
form workable ores.[3]

SOILS—The upper layer of the regolith, the *soil*, is, perhaps,
the most important residual deposit.   It grades downward
into loose, broken, rock debris, the *subsoil*, which, in its turn,
passes into the solid bedrock.   Soils may form on the top of
either residual or transported materials.   In the first case,
soil and subsoil are of like nature to the bedrock; in the
second, the detrital beds may be regarded as the subsoil, and
both soil and subsoil may be composed of materials foreign
to the bedrock.   Hence a distinction arises between *residual*
or *sedentary*, and *transported* soils.

In the agricultural sense soils are the superficial layers,
usually less than a foot in thickness, of disintegrated and decomposed rock material, which is mingled with organic
matter, and furnishes the necessary conditions and materials
for plant growth.   Soils are thus composed of mineral and
organic substances with the former usually in great predominance, although in peaty and swampy soils the organic
matter may reach to over 50 per cent. of the whole.   As
residual materials soils consist of stones, sand, silt, and clay,
in various proportions, and their textures are determined
by the relative amounts of these constituents.   The stones
may be derived from the bedrock, or they may have been
transported from distant localities, and may be of many

[1] *Geol. Mag.*, 1903, p. 63.
[2] *Mining Mag.*, 17, 1917, pp. 67-77; 120-8; 171-9; 220-9.
[3] W. G. Miller, " Lateritic Ore Deposits," *Rept. Ont. Bur. Mines*,
xxvi, pt. 1, 1917, 19 pp.

different varieties. The sand grade may consist of any of the common rock-forming minerals with a decided predominance of the more stable types such as quartz, white mica, and some of the rarer heavy minerals. The silt and clay consist partly of rock and mineral flour, and partly of the hydrous aluminium silicates, ferric and aluminous oxides and hydrates, which are the products of rock decomposition. Calcium carbonate of sand or clay grade is an important constituent in some soils.

The organic material in the soil is *humus*, derived from the bacterial decay of the entombed vegetable and animal matter. It provides, on the one hand, plant-food in a readily assimilable form ; and, on the other hand, it yields organic acids by continued decomposition, which help to dissolve some of the mineral constituents, and thus render available other essential plant-foods. Agricultural investigators classify soil constituents as sand, silt, clay, calcium carbonate (" lime "), and humus. According to the degree of admixture of these constituents, such types as sandy soils, loams (sand + silt or clay), marls (clay or silt + calcium carbonate), silty soils, clay soils, calcareous soils, and peaty soils, may be distinguished. Sandy, loamy, and calcareous soils are light, dry, friable, and porous, and are apt to be more or less depleted of soluble plant-foods ; clay soils, on the other hand, tend to be heavy, dense, impervious, and wet, but retain the soluble plant-foods better than the sandy soils.

By modern agricultural chemists the soil is regarded as a colloidal system (p. 219) in which the essential constituents are colloidal clay and colloidal organic matter.[1] It is the presence of these substances that distinguishes soil from mere finely-divided rock material. Colloidal clay or the *clay complex* is the substance which confers the property of plasticity and other characteristics on soils and natural clays. It is a material of indefinite composition containing silicic acid and the hydrated oxides of aluminium and iron. In association with organic matter in the same physical condition it forms a gel-coating or film on the surfaces of the more or less decomposed mineral particles of the soil, and

[1] The following summary has been prepared from a paper by G. W. Robinson on " Pedology (Science of Soils) as a Branch of Geology," *Geol. Mag.*, lxi, 1924, pp. 444-55.

possesses the property of *adsorption;* i.e. it is able to effect a concentration of, and retain, certain substances such as calcium and potassium salts, when solutions containing these materials come into contact with it. As these and other adsorbable substances are essential plant-foods, the importance of the colloidal matter can hardly be over-estimated.

To sum up, the soil consists of more or less decomposed rock and mineral particles of varying sizes, associated with a hydrated colloidal complex of clay and organic matter, which may be conceived as a matrix surrounding the particles, and facilitating their aggregation into compound particles or crumbs. Variations in the relative amounts of these constituents, and in the grain-size, account for all the different kinds of soil. Sandy soils consist of coarse mineral particles with a comparatively scanty matrix of colloidal matter ; clay soils consist mainly of the finest rock and mineral flour, with correspondingly larger quantities of material in the colloidal condition. In a raw subsoil clay the colloidal matter is inorganic, whilst in a surface soil, especially when cultivated, there is a comparatively large amount of colloidal organic matter as well.[1]

[1] For a modern study of a particular soil on these lines, see H. B. Milner, " Paraffin Dirt," *Mining Mag.*, xxxii, 1925, pp. 73-85.

CHAPTER XI

SEDIMENTARY ROCKS; MINERALOGICAL, TEXTURAL, AND STRUCTURAL CHARACTERS

THE *sedimentary* or *detrital* rocks are those formed by the deposition of the solid materials carried in suspension by the agencies of transport. Their mineralogical, textural, and structural characters, dealt with in the present chapter, afford evidence as to the conditions attending the deposition of the material, the extent of denudation, the mode of transport, climate, and other palæogeographical features of the period to which the rocks belong.

MINERAL COMPOSITION—The minerals of sedimentary rocks fall into two classes: (1) the insoluble residues of rock decomposition; and (2) the comparatively durable minerals derived from pre-existing rocks. Amongst the former are the groups of (a) the clay minerals, such as kaolinite, halloysite, etc.; (b) the micaceous minerals, including the hydromicas and chlorite; (c) the aluminium hydroxides, bauxite, gibbsite, etc. (see p. 183); and (d) the ferric oxides and hydroxides. Minerals of a ferro-magnesian group, such as serpentine and talc, are rarely found. A large number of mineral species are found in the second class: quartz is, of course, always the most abundant, and certain accessory minerals, such as zircon, rutile, tourmaline, garnet, kyanite, magnetite, etc., are frequently present in sediments. With rapid disintegration, however, any rock-forming mineral, even those most susceptible to decomposition, may for a time form a constituent of a sedimentary rock. Thus fresh olivine may occasionally be found in the shore sands of the Ayrshire coast, having been derived from the rapid breaking-down of the numerous olivine-rich rocks of that county.

The mineral composition of a sediment is affected by many factors. Since the insoluble residues of rock decomposition

187

are usually of flour-like fineness they form the chief constituents of clays; whereas sands consist chiefly of quartz and
felspars, since these minerals are the most abundant of the
relatively durable minerals, and are derived from the disintegration of rocks the grain-sizes of which are comparable to
those of sands.

The mineral composition also depends on the nature of the
rocks forming the gathering-ground of the material. If the
country-rock consists mainly of some mineralogically uniform
rock, such as quartzite, or a granite poor in ferromagnesian
minerals and accessory minerals, the composition of the
sediment resulting from its denudation will be simpler than
that resulting from the waste of a lithologically or mineralogically heterogeneous region.

The duration and nature of the transport is also a factor in
determining the mineral composition. Long-continued drift-
ing to and fro by wind or by water tends to effect a separation
of particles according to mass and surface area, and, therefore, according to composition. Wind is a particularly
efficient sorter of sand grains. In deserts, mica flakes and
dust are blown far away, and the remaining sands are sifted
and re-distributed until there is an approach to mineral
uniformity. Long-continued transport in rivers or along
shores may be almost equally effective in producing clean,
graded, and uniform deposits. The process of panning or
washing for gold or other heavy constituents from sands or
gravels is an illustration of the effectiveness of these modes
of sorting. The streaks and patches of sands rich in garnet
or magnetite, found on some beaches, are examples of
natural panning.

The vicissitudes of transport tend to destroy the softer,
more cleavable, and brittle mineral grains, and thus to produce greater mineral uniformity in the final material. Dr. W.
Mackie has shown that the proportions of felspars in the
sands of the rivers Spey and Findhorn become progressively
smaller towards the mouths of those rivers.[1] The felspars
are more altered and softer, and are consequently more easily
broken down to silt and clay grade than is the hard and
intractable quartz. The waves of the sea are still more

[1] "Sands and Sandstones of Eastern Moray," *Trans. Edin. Geol.
Soc.*, 7, 1899, p. 149.

effective in eliminating the felspars from sands; for, according to Dr. Mackie, the average amount of felspar in the sands of the Moray rivers at points near the sea is 18 per cent. whereas, in the adjacent shore sands, the proportion is only 10 per cent. Other minerals may be similarly eliminated from sands in proportion to their softness, susceptibility to alteration, and cleavability.

Those constituents, such as boulders, pebbles, or mineral grains, which have been formed elsewhere, and have been brought into a sediment from outside, are termed *allogenic* (= originating elsewhere); those constituents which have been formed *de novo* within the sediment by subsequent changes are called *authigenic* (= formed in place or on the spot).

GRAIN SIZE—The principal factors which determine the size of grain in sedimentary rocks are the dominant modes of weathering (decomposition or disintegration), the texture and composition of the pre-existing rocks, and the kind and amount of the transport suffered by the material.

The effect of decomposition in general is to produce material of flour-like fineness. Disintegration, on the other hand, tends to produce detritus of coarser grain, since it is operative along cleavages, and around the margins of crystals, in coarse-grained rocks, and along the joints and fissures of fine-grained rocks. Thus the material produced is either approximately of the same grain size as the parent rock, or forms larger fragments the size of which is only limited by the spacing of the fissures. The texture, composition, and jointing of the pre-existing rock thus become factors in the grain-size of the resulting sediments, and operate in a rather paradoxical way. A coarse rock such as granite will give rise to a mass of broken crystals of nearly the same grain size; but a dense rock, such as felsite or compact limestone, will be disrupted along its fracture planes, and the fragments will therefore tend to be larger than those derived from the coarser rock.

The amount of transport undergone by the detritus will obviously affect the grain size of the resulting sediment. A longer period of shifting to and fro by wind, waves, rivers, or glaciers, inevitably means a greater amount of wear on each particle, and consequently finer grain. The impact of particles upon each other or upon rock surfaces is heavier in air than in water; hence æolian deposits are apt to be finer than aqueous deposits which have suffered an equivalent amount of transport. A further discussion of this subject is given under the head of Rounding (p. 193).

The terms applied to detrital material of varying sizes, such as boulders, cobbles, pebbles, gravel, sand, silt, clay, are rather loosely defined. Recent attempts to define them have been made by Grabau,[1] and Boswell.[2] The following table represents a compromise between the various systems, and in the main follows that given by Holmes.[3] Materials between the various arbitrary limits of size are known as *grades*.

Grade.	Range of Size of Fragments.	Main Groups.
Boulders .	Greater than 200 mm. (about 8″) (least diam.)	⎱ Rudytes.
Cobbles .	Between 200 mm. (8″) and 50 mm. (2″)	
Pebbles .	,, 50 ,, (2″) ,, 10 ,, (½″)	
Gravel .	,, 10 ,, (½″) ,, 2 ,, ($\frac{1}{12}$″)	Gravel.
Very coarse sand .	,, 2 ,, ($\frac{1}{12}$″) ,, 1 ,, ($\frac{1}{25}$″)	⎱ Sand.
Coarse sand	,, 1 ,, ($\frac{1}{25}$″) ,, ·5 ,, ($\frac{1}{50}$″)	
Medium sand	,, ·5 ,, ($\frac{1}{50}$″) ,, ·25 ,, ($\frac{1}{100}$″)	
Fine sand .	,, ·25 ,, ($\frac{1}{100}$″) ,, ·1 ,, ($\frac{1}{250}$″)	
Silt . .	,, ·1 ,, ($\frac{1}{250}$″) ,, ·01 ,, ($\frac{1}{2500}$″)	Silt.
Dust, mud, clay .	Less than ·01 mm. ($\frac{1}{2500}$″)	Clay.

Sediments which contain a large number of grades in more or less equal amount are said to be unassorted or ill-assorted. On the other hand, sediments containing a large proportion of one grade are said to be well-assorted or graded. Ill-assorted sediments are produced under conditions of rapid and confused deposition (as e.g. glacial deposits), when fragments of the most varied sizes are mixed together. Well-sorted sediments, on the other hand, are produced by long-continued transport in wind or water, with the consequent sifting out of particles of differing sizes.

[1] *Principles of Stratigraphy*, 1913, p. 287.
[2] *Memoir on British Resources of Refractory Sands*, 1918, p. 13.
[3] *Petrographic Methods and Calculations*, 1921, p. 197.

The proportions of the various grades present in a sediment are represented by its mechanical analysis. Sifting, elutriation, or subsidence methods are used in ascertaining the mechanical composition of a sediment, but their description is beyond the scope of this work.[1]

Professor P. G. H. Boswell has described an ingenious method of graphically expressing the results of the mechanical analysis of a sediment by means of curves.[2] The grade sizes are plotted horizontally, the scale being proportional to their logarithmic values; and the cumulative percentages by weight of material above the various grade sizes are set off vertically (Fig. 53). Thus the London clay represented by the curve LC has 17 per cent. of material of diameter greater than 0·1 mm., and about 55 per cent. of diameter between 0·1 mm. and ·01 mm. Hence the ordinate at grade 0·1 mm. is 17, and at ·01 mm. is $17 + 55 = 72$. Horizontality in any part of a curve means the absence of the grade corresponding to that part of the abscissa over which the horizontal part extends. Thus in the curve BB (sand of Bovey Beds) the silt and clay grades are practically absent. Verticality means a considerable percentage of the grade represented by that part of the curve. Thus a pure gravel, consisting entirely of grains not less than 2 mm. in diameter, is represented by the vertical line GG; a pure sand of medium grade is shown by the line SD, and pure silt and clay by lines such as ST and CL. A well-assorted sediment, then, is indicated by approximate verticality in a considerable portion of its curve; an ill-assorted sediment is indicated by diagonally arranged curves (PB and CS). This graphical mode of representation is of great value, as it shows how closely natural sands, for example, approximate to the ideals for certain economic purposes. Thus the curves MM and GS represent the mechanical analyses of an ideal moulding sand and an ideal glass sand respectively.

HEAVY MINERALS IN SANDS AND SANDSTONES—In sand accumulation by whatever mode of transport the assorting of constituents of differing composition depends chiefly on the

[1] For methods see P. G. H. Boswell, *Memoir on British Resources of Refractory Sands*, 1918, pp. 1S 28; or A. Holmes, *Petrographic Methods and Calculations*, 1921, pp. 204-15.
[2] P. G. H. Boswell, *Proc. Geol. Assoc.*, 27, pt. 2, 1916, p. 98.

mass of the particles. Grains of the same mass tend to be concentrated. Other things being equal, particles of a certain size and specific gravity will be concentrated along with particles of smaller size but greater specific gravity. As quartz and felspar are by far the most abundant constituents of sands, and are of nearly the same specific gravity.

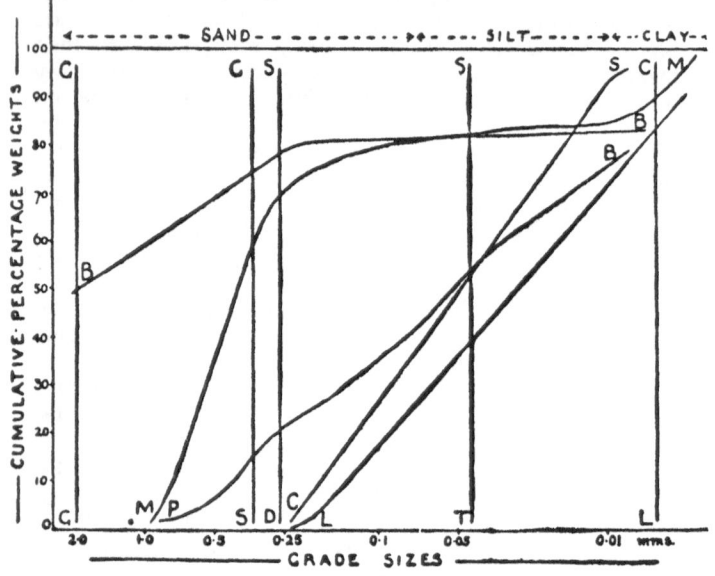

FIG. 53.—GRADE SIZE OF SEDIMENTS.
Graph illustrating the variation in grade size of a number of sediments.
(For explanation see text, p. 191.)

long-continued transport generally leads to a comparative uniformity in size of these fragments; but they will be associated with a number of heavier minerals which, in general, will be of smaller sizes. The heavy minerals are, therefore, mostly to be found in the finer grades of sand, and are often practically absent from the coarser grades.[1]

[1] V. C. Illing, in P. G. H. Boswell: *Memoir on British Resources of Refractory Sands*, 1918, p. 180.

The nature and relative amounts of these crops of heavy minerals in different geological formations vary with the different conditions of deposition, and with skilled inter-pretation they can be made to yield valuable data regard-ing the sources of the containing sediments, and of the geographical conditions attending their deposition. The light crop of quartz and felspars, too, yields much informa-tion as to sources, climate, and palæogeography.

The separation of the light and heavy mineral crops if sands is effected by the use of bromoform (specific gravity, 2·8) in a funnel. The quartz and felspar fragments float, whereas the heavy minerals sink and can be drawn off through a filter. The proportion of heavy minerals is very small. Usually it is much less than 1 per cent., and only very exceptionally does it reach as much as 5 per cent. Ilmenite, magnetite, zircon, rutile, tourmaline, and garnet, are amongst the commonest minerals of the heavy suite, and appear in nearly every British geological formation of arenaceous com-position. Other heavy minerals may be locally abundant. The heavy mineral crop may be diagnostic for a particular formation if the rarer minerals are noted, as well as the crystallographical and optical peculiarities of the commoner ones.[1]

SHAPE AND ROUNDING OF GRAINS (Fig. 54)—In the clastic rocks the shapes of the constituent fragments depend on the original shapes of the material supplied by weathering, and on the amount and kind of transport it has suffered. The first factor depends on the physical characters of the dif-ferent minerals concerned. The quartz of a granite will supply sharply-angular fragments ; the felspars will provide grains bounded by the parallel cleavage fractures ; the micas will give rise to thin, irregularly-bounded flakes ; zircon and apatite may be liberated as perfectly-shaped crystals. In the case of a fine-grained compact rock, the fragments pro-duced by disintegration will generally be angular, as they will be determined by the intersection of natural fissures, or of irregular fracture planes due to the mode of disruption.

---

[1] See P. G. H. Boswell, " The Petrography of the Sands of the Upper Lias and Lower Inferior Oolite in the West of England," *Geol. Mag.*, lxi, 1924, p. 258;

It may be said that, normally, the material supplied by weathering is initially angular.

The matter may be complicated by the fact that rocks the fragments of which have been shaped during a previous cycle

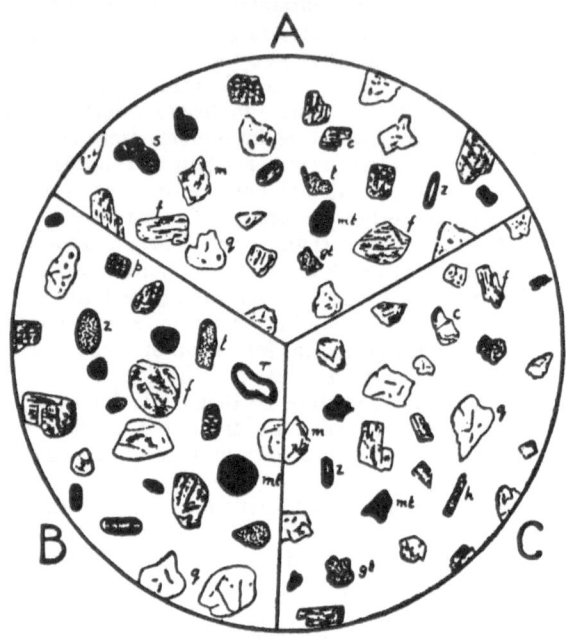

Fig. 54.—Shape and Rounding of Sand Grains.
A. Shore sand, Ayr.   Most of the grains subangular.
B. Desert sand, Assouan, Egypt.   Many of the grains perfectly rounded.
C. Fluvioglacial sand, Glen Fruin, Dumbartonshire.   The grains sharply
       angular.
       q, quartz; *f*, felspar; *m*, mica; *c*, chlorite; *p*, pyroxene;
h, hornblende; *s*, zircon; *t*, tourmaline; *r*, rutile; *gt*, garnet;
mt, magnetite and iron ores; *s*, serpentine.
   Magnification about 20 diameters.

of denudation may occur in the area undergoing waste. Thus, by disintegration, desert sandstones, or loosely-compacted oolites, would from the beginning supply perfectly rounded grains. Schists, again, would provide elongated, prismatic, or lenticular fragments which, on rounding, would become cylindrical or spindle-shaped.

The effect of transport upon the initially angular mineral and rock fragments supplied by weathering is to smooth and round off irregularities, and finally to produce rounded and polished grains. The degree of rounding depends on size; the largest particles are the most rounded. It depends also on weight; size for size the specifically heavier particles suffer the most rounding. Hardness militates against rounding; the softer a mineral fragment the more likely is it to be well rounded. Thus felspar grains frequently show a greater degree of rounding than quartz. Another factor is the distance of transport. Other things being equal, the longer the travel the greater the rounding. Finally, rounding depends on the agent of transport. For a given distance of travel, particles carried by ice are the least, and those carried by wind are the most, rounded; whereas those transported by water occupy an intermediate position in this respect.

The nature of the medium of transport affects rounding in another way depending on its viscosity.[1] When a grain is reduced below a certain size in water, impact with other grains, or with the river-bed, becomes impossible owing to the repulsive force of surface tension exerted by the thin skin of water with which the grain is surrounded. Ziegler believes it improbable that grains of diameters less than ·75 mm. can be well rounded in water.[2] All grains of smaller sizes than this will remain angular, and the transition in size between angular and rounded particles will be abrupt. The repulsive force of surface tension is much less in air than in water, and impact is therefore possible between grains of much smaller diameters than ·75 mm. Hence very small grains may be worn down and rounded by transport in air, and it is possible by this means, in conjunction with other features, to distinguish between sand deposits of aqueous and æolian origin.

COHESION—When first formed sediments are loose, soft, and unconsolidated, but in course of time they become firm, hard, and compacted. This is due chiefly to two processes:

[1] J. G. Goodchild, " Desert Conditions in Britain," *Trans. Edin. Geol. Soc.*, 7, 1909, p. 301.
[2] V. Ziegler, *Journ. Geol.*, 19, 1911, p. 645.

(1) welding or induration; (2) cementation. *Welding* is consolidation by pressure due either to the weight of super-incumbent material, or to earth movement. Most of the water in the sediment is squeezed out, and the particles cohere because they are brought within the limits of mutual molecular attraction. The fine-grained clay rocks are more susceptible to this mode of consolidation than the coarser sand rocks. The process is imitated by means of hydrostatic pressure in the manufacture of many artificial materials.

In *cementation* the particles are bound together by the deposition of cementing substances between the grains. These substances may be brought into the rock by percolating solutions which may carry silica, calcium or magnesium carbonates, or iron salts. The first two are deposited between the grains just as mortar between bricks; the third may be deposited likewise, but often iron oxide forms a thin film around each grain. It is then a very effective bond, especially if it is in the hydrated colloidal condition. The cement, again, may be clayey matter which has been produced by the decomposition of felspar within the sediment, or which has been deposited along with the predominant sand grains. Micaceous substances due to original deposition, or to subsequent decomposition processes, may also act as cements. In most cases, therefore, cementing substances are authigenic.

The processes of welding and cementation generally act together in the hardening and consolidation of a sediment. Thus dust, mud, or clay is converted into clay, mudstone, or shale; sand is changed into sandstone; pebbles, cobbles, and boulders are cemented together to form conglomerates and breccias.

STRATIFICATION [1]—One of the most characteristic features of sedimentary rocks is their deposition in beds, layers, or strata. The *bedding* or *stratification* is indicated by differences of composition, texture, hardness, cohesion, or colour, disposed in approximately parallel bands. Stratification, otherwise obscure, is frequently brought out clearly by weathering processes. The plane of junction between different beds is a *bedding-plane*. Not infrequently it is also a

----

[1] A recent discussion of stratification is given by K. Andrée, *Geol. Rundschau*, vi, 1915, pp. 351-98.

plane of discontinuity, along which the rock splits more or less easily.

If loose material of varying grades is dropped into still water, the largest pebbles and sand grains will be found on the floor of the basin of deposition, and in upward succession there will be found progressively finer material with the finest of all on top. A regular stratification according to size from bottom to top will thus be produced (Fig. 55 A). If however, the medium of transport is a current moving in a particular direction, then the gradation takes place, not in a vertical, but in a horizontal, direction. The coarsest material is deposited first, then successively finer and finer sediment,

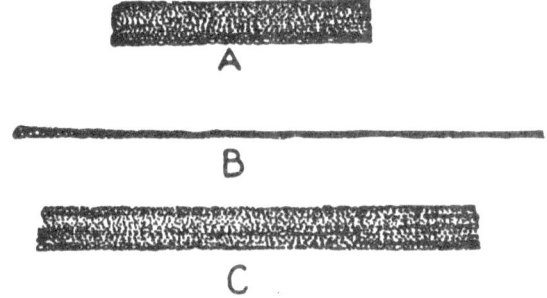

FIG. 55.—DIAGRAM ILLUSTRATING MODES OF STRATIFICATION.
(For explanation, see text, p. 197.)

until the finest clay is dropped at the remotest point from the source (Fig. 55 B). The sediment is then graded horizontally but not stratified. If, however, the process is repeated again and again with a current of continually varying velocity, new layers of differing grades will be laid down on the old, and stratification will result (Fig. 55 C).

A single layer bounded by two bedding planes is a *bed* or *stratum*. The thicknesses of beds may vary from many feet down to a fraction of an inch. Very fine, paper-thin beds are known as *laminæ*, and are found only in material of clay, silt, or very fine sand grade. Lamination is often due to the deposition of minutely-platy minerals such as the micas. It may also be brought about by pressure due to an overlying load acting upon minute rods, scales, and flakes, rotating them into a position perpendicular to the pressure direction. Stretching or flow of the fine material may cause the same ultimate result. The orientation of the constituents in this way brings about *fissility*, the ability of certain fine-grained sedimentary rocks to split easily along the bedding planes.[1] Fissility in coarser rocks may be due to the intercalation of thin layers of clayey or micaceous material, as in many thin-bedded sandstones (faikes).

A bedding-plane or lamina frequently indicates a longer or shorter period of non-deposition, especially when it separates materials of appreciably different composition or texture, for the differences of conditions necessary to produce these layers cannot be established at once. A collection of beds forming a definite unit is a geological formation.

When the bedding-planes are disposed approximately parallel to one another, the phenomenon may be called *concordant* bedding. Often, however, there may be seen within particular beds a subsidiary stratification indicated by bedding-planes which are inclined to the major lines of bedding. These appearances, which are variously known as cross-, oblique-, false-, or current-bedding, may all be comprised under the term *discordant* bedding.

Several kinds of discordant bedding have been distinguished by Andrée (*op cit.*). The most important, perhaps, is *current* bedding. In this type beds with oblique lines of stratification are seen bounded by layers of concordant bedding (Fig. 56). Current-bedding indicates rapid changes in direction and strength of a stream of water carrying sediment, in bars, spits, sand-banks, and deltas. The mode of deposition is the same as that by which a railway embankment or a tip-heap is built. The sediment is carried horizontally by the current and dropped over the terminal slope.

The structure of deltas well illustrates the production of current bedding on both a large and small scale. There are three sets of beds ; topset, foreset, and bottomset (Fig. 57). The topset beds are those which are laid down on the subærial surface of the delta with a dip equal to the original slope of that surface ; the foreset beds are those built by ordinary

[1] For a discussion of fissility and its causes see J. V Lewis, *Bull. Amer. Geol. Soc.*, 35, 1924, pp. 557-90.

current action in the water ; the bottomset beds consist of fine mud or silt floated out and deposited on the floor of the basin of deposition in advance of the foreset beds.

Torrential bedding shows an alternation of coarse, current-bedded materials, and fine horizontal laminæ

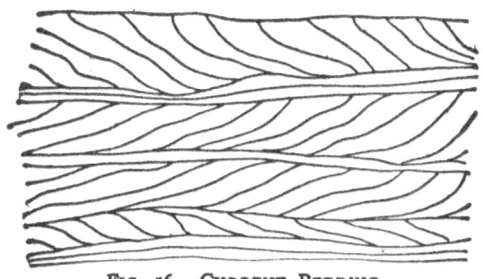

FIG. 56.—CURRENT BEDDING.
(For explanation see text, p. 198.)

(Fig. 58). It is found in alluvial fans built in semi-arid regions by streams which, in times of flood, push forward a load of coarse material, and at quieter times deposit only fine clays or silts.

Æolian current bedding, due to wind, shows curved cross lamination of larger scale and wider radius than that

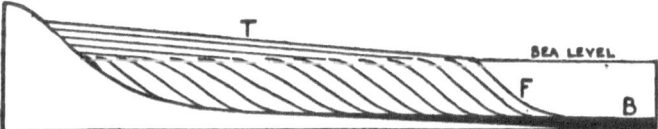

FIG. 57.—DELTA STRUCTURE.
Diagram illustrating structure of deltas. T, topset beds ; F foreset beds ; B, bottomset beds. (See text, p. 198.)

due to water. It represents the remnants of successive dunes which have marched across the region in the direction of the prevalent wind. It is also marked by extreme irregularity owing to frequent alternations of deposition and denudation under the influence of air currents of varying direction and strength.

Minor Structures and Markings [1]—The well-known wavy pattern, or *ripple-mark*, so often seen on sandy beaches, may, under special conditions, be preserved in the sediments. On beaches the ripple-marks are due to wave action, which produces miniature undulations in the sand having symmetrical cross-sections (Fig. 59 A).

An unsymmetrical type of ripple-mark, however, is made by current action either in water or air (Fig. 59 B). The sand particles are swept up the long slope of the ripple, pushed over the crest, and dropped into the trough ; and the ripples thus slowly migrate like dunes.   In æolian ripple-marks the coarser grains are found on the crests, and the

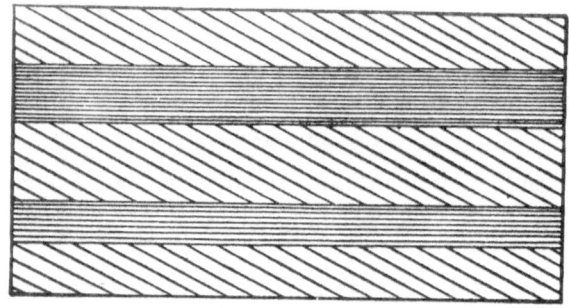

Fig. 58.—Torrential Bedding.
(For explanation see text, p. 199.)

finer ones in the troughs, of the undulations.   In aqueous ripple-marks, on the contrary, the finer grains are on the crests, while the coarser ones are found in the troughs and on the short, steep slopes of the ripples.

Mudcracks or *suncracks* such as can be seen on the floor of any dried-up pool, are often preserved in the fine-grained

[1] The following papers contain recent discussions of these features : " Ripple-marks," E. M. Kindle, *Geol. Surv. Canada, Mus. Bull.*, 25, 1917, pp. 121 ; W. H. Bucher, *Amer. Journ. Sci.*, 47, 1919, pp. 149-210 ; " Mudcracks," E. M. Kindle, *Journ. Geol.*, 25, 1917, pp. 135-44 ; " Rain-prints," etc., W. H. Twenhofel, *Bull. Geol. Soc. Amer.*, 32, 1921, pp. 359-72 ; " Tracks and Trails," P. E. Raymond, *Amer. Journ. Sci.*, 3, 1922, pp. 108-14.   For a fine illustration of the track made by a dying lobster, see F. A. Bather, *Knowledge*, Sept., 1914, p. 329, and Fig. 325.

sedimentary rocks. They form a network of fissures enclosing polygonal areas, and may be preserved by infilling with sand, or a different kind of mud. They are indicative of conditions involving prolonged exposure of clayey sediments to the atmosphere, and are therefore characteristic of the flood-plains of large rivers, although they may indicate, much more rarely, tidal mud-flats or shelving lake shores.

Rain-prints may also be found on the surfaces of sedimentary rocks, and are preserved in the same places, and under the same conditions, as mudcracks. A rainprint is a slight shallow depression encircled by a low ridge which is raised by the impact of the drop. If the rain falls obliquely, the ridge on the lee side is higher than that on the windward

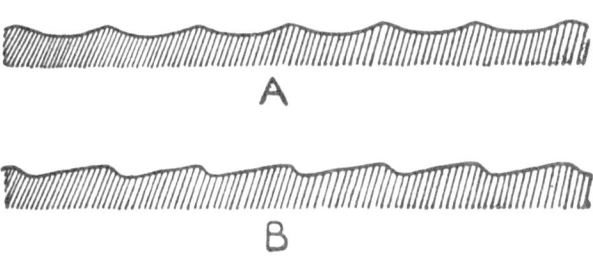

FIG. 59.—RIPPLE MARKS.
Vertical sections through ripple marks. A, wave ripple marks.
B, current ripple marks. (See text, p. 200.)

side. Twenhofel (*op. cit.*) has shown by experiment that impressions resembling rainprints may be produced in many different ways: (1) Real rainprints; (2) Hail impressions; (3) Drip impressions; (4) Spray and splash impressions; (5) Impressions due to bubbles produced in various ways; (6) Pit and mound structures due to small upward currents produced in rapidly flocculated sediments. Only a small percentage of the impressions usually described as rainprints are due to falling rain. Hence caution has to be exercised in interpretations of conditions based on these structures.

Tracks and Trails—These are markings indicating the passage of some animal over soft sediment which was able to take and retain the impression. Footprints due to amphibia, reptiles, and birds have often been observed in the strata. Long, winding, shallow, parallel-sided depressions, are usually ascribed to worms. Raymond, however, has given evidence to show that the trails on sandstone and shale, usually attributed to worms, are really made by gastropods and other short-bodied animals. Winding, irregular trails with sharp turnings, must have been made by short, not elongated, animals. The trails of the common earthworm are nearly straight, or form curves of long radius.

CHAPTER XII

SEDIMENTARY ROCKS—DESCRIPTIVE

CLASSIFICATION—The factor of grade size is the one most used for the classification of sedimentary rocks (p. 190). Four groups may be distinguished :—

1. *Rudaceous* [1]—Rocks consisting chiefly of gravel, pebbles, cobbles, or boulders. Loose materials of this class are: gravels, pebble-beds, shingle, boulder-beds, scree, talus, etc. When cemented they form conglomerates and breccias.

2. *Arenaceous*—Rocks consisting chiefly of material of sand grade. Loose materials are sands ; when consolidated they form sandstones, grits, arkoses, graywackes, etc.

3. *Silt Rocks*—Rocks consisting chiefly of material of silt grade. Silt and siltstones. This class is usually included either with the arenaceous or the argillaceous classes, but several distinct rock types occur in it, and the group is worthy of a separate designation.

4. *Argillaceous*—The clay rocks, consisting of the finest materials of rock decay. Dust, mud, clay, when more or less unconsolidated ; mudstone and shale when compacted.

This grouping is also partly chemical and mineralogical. From the rudaceous to the argillaceous rocks an increasing definiteness of chemical and mineral composition can be traced. The pebble- and boulder-rocks form the most heterogeneous group from the point of view of composition. The arenaceous rocks, however, composed mostly of quartz with subordinate felspar, are necessarily highly siliceous. The argillaceous rocks, which comprise the finest products of rock decay, and especially of decomposition, are consequently made up chiefly of hydrous aluminium silicates. Silt rocks are intermediate in composition between the arenaceous and

[1] Grabau, *Principles of Stratigraphy*, 1913, p. 285.

the argillaceous groups. Nevertheless, the composition of each group may vary between wide limits. Even in the argillaceous group rocks devoid of hydrous aluminium silicates may occur, as, e.g. a detrital limestone mud.

Another mode of grouping the sedimentary rocks is according to the agent most prominent in producing their present characters, as aqueous (water-formed), æolian (wind-formed), glacial (ice-formed). Still another mode is according to the location of deposition, as marine, continental, fluviatile, estuarine, lacustrine, etc. Both these modes of grouping, however, are also applicable to the other classes of secondary rocks.

RUDACEOUS ROCKS.[1]   BRECCIAS[2] AND CONGLOMERATES—A fundamental distinction between angular and rounded frag-

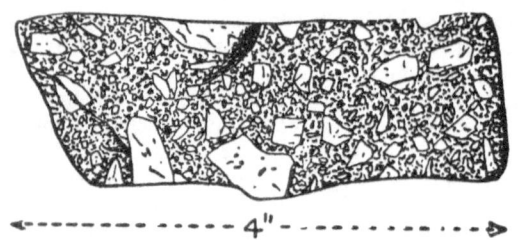

FIG. 60.—BRECCIA.

Quartz-breccia (Cambrian), Inchnadampff, Sutherlandshire. Shows angular fragments of white quartz in coarse, gritty matrix.

ments, depending on the amount of transport the material has suffered, is the foundation of subdivision in this group between scree, talus, and breccia, on the one hand, and shingle and conglomerate on the other. In the first class the material has suffered little or no transport ; it has simply accumulated at the foot of a slope. In the second class the fragments have undergone considerable transport in water, and have consequently become more or less rounded. The angular rudaceous rocks, or *breccias* (Fig. 60), are very hetero-

[1] For short accounts of some British rudaceous rocks see A. Holmes, *Petr. Methods and Calc.*, 1921, pp. 168-74.
[2] W. H. Norton (*Journ. Geol.*, 25, 1917, pp 160-94) gives a useful classification and general account of breccias. See also S. H. Reynolds, *Geol. Mag.*, lxv, 1928, pp. 87-107.

geneous in composition ; the fragments may consist of any type of rock which has undergone subaërial weathering. On the other hand, *conglomerates* consist only of the more durable fragments which, by reason of their hardness and toughness, have been enabled to survive the vicissitudes of transport, and in doing so have been worn down and rounded.

The huge accumulations of *talus* or *scree* which occur at the bases of slopes in semi-arid, desert, or polar regions, where disintegration is the chief mode of weathering, are examples of loose rudaceous deposits of angular materials.[1] H. T. Ferrar [2] has described material of this kind partially filling the wadis or steep-sided gorges in the region between the Nile and the Red Sea. It is very fresh and sharply angular, and in their short transport by the infrequent desert floods the blocks are often scratched, grooved, and shattered. These accumulations are very similar to the " trappoid " breccias of the Upper Permian in the English Midlands, which consist of angular blocks of volcanic rocks, grits, slates, and limestones, identical with, and doubtless derived from, the pre-Permian rocks to the west. The *brockram* of the Cumberland Permian is a similar type of breccia, in which the fragments are mainly of Carboniferous limestone.

While breccias are mainly formed under subaërial conditions, they may infrequently be of marine origin. The outstanding example is that described by M. Macgregor from the Jurassic of Helmsdale, Sutherlandshire.[3] These breccias consist of angular blocks of Old Red Sandstone, of all sizes up to a mass measuring 150' × 90' × 30' (probably a fallen stack), embedded in fossiliferous marine Jurassic muds, which are contorted around the included blocks. These deposits represent talus accumulated at the bases of Old Red Sandstone cliffs which bordered the Jurassic sea. The conditions which must have obtained at their formation are now being repeated along the present Caithness coast. Other types of breccia are those formed by the disintegrative action of volcanic explosion. A volcanic breccia is called *agglomerate*. It is composed chiefly of igneous materials, but may contain

[1] A. C. Lawson (*Bull. Dept. Geol. Univ. Cal.*, Publ. vol. vii, 1913) proposes the term *fanglomerate* for this material when it is spread out as alluvial fans.
[2] *Rept. Brit. Assoc. Australia*, 1914, p. 362.
[3] *Trans. Geol. Soc Glasgow*, vol. xvi, 1916, p. 75.

non-volcanic fragments brought up from the basement of the volcano, or torn from the sides of the vent. Agglomerate is heaped up as a tumultuous mass of blocks of all sizes in the vicinity of the eruptive orifice.

Breccias may also be formed along planes of movement in the crust. These *fault-* or *crush-breccias* represent the first stage of mylonisation (p. 284) or milling of rocks.

Intraformational breccia is made by the cracking and breaking up of a clay or silt under the sun's heat or by drying, and the incorporation of the angular fragments in a succeeding bed of sand.

Loose rudaceous deposits with rounded fragments are described as *gravel*, *pebble-beds*, or *shingle*. Gravel consists of more or less rounded fragments of diameters between 2 mm. and 10 mm.; pebble-beds and shingle (practically synonymous terms) consist of larger fragments ranging up to 50 mm., or 2 inches in diameter. Gravel and shingle are composed of hard, tough, durable materials which are able to stand a great deal of wear. The extensive gravel deposits resting on the Chalk of the South of England, and the great shingle beaches of Dorset (Chesil Beach), and of Kent (Dungeness), are thus mainly composed of flint, the most abundant hard rock of that region. Some of the gravel sheets were probably in the first instance spread as outwash fans in front of the Pleistocene ice sheet, but have been partly re-arranged by rivers. The shingle beaches represent pebble-beds of marine origin, which have been formed by continual wave and current transport from west to east along the south coast.

Cobble- and *boulder-beds* are formed by torrential action where large rivers debouch from mountain valleys on to an adjacent lowland. They are built up as alluvial fans at the mouth of each valley, and in course of time may become confluent one with another along a mountain front. In this way formations many hundreds or even thousands of feet thick are now being produced along the margins of the greater mountain ranges.

Owing to the inability of sea waves and currents to shift coarse materials at more than shallow depths, marine pebble-beds are limited to less than 100 feet of thickness.[1] Thick

[1] J. Barrell, *Bull. Geol. Soc. Amer.*, 36, 1925, pp. 279-342; H. E. Gregory, *Amer. Journ. Sci.*, 39, 1915, p. 487.

and coarse pebble- and boulder-beds are, therefore, most probably of fluviatile origin. The great conglomerates of the Lower Old Red Sandstone in Scotland, built chiefly of quartz and quartzite boulders derived from the Highlands, are interpreted in this sense. The Torridonian conglomerates of the north-west Highlands, and the Budleigh Salterton (Devonshire) pebble-beds of Triassic age, are also regarded as of fluviatile origin. Undoubted marine conglomerates are usually found as thin beds succeeding an unconformity,

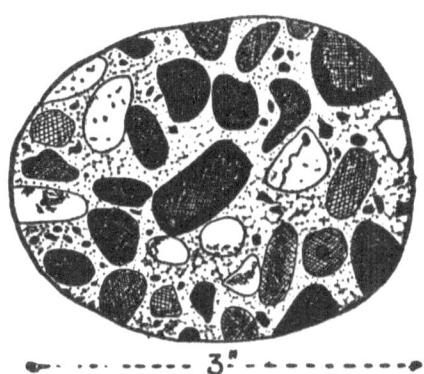

FIG. 61.—CONGLOMERATE.

Cut and polished pebble of flint-conglomerate (Hertfordshire puddingstone). Shows rounded and variously-coloured flint pebbles in fine-grained siliceous matrix, which also contains angular chips of black flint.

and are due to the gradual re-advance of the sea during a subsidence.

For petrographic purposes the Hertfordshire Puddingstone, consisting of rounded flint pebbles in a matrix of quartz fragments, which have been cemented into a hard rock by the infiltration of silica, may serve as a typical example of a *conglomerate* (Fig. 61). It is a local facies of the Reading pebble-beds (Eocene), which is found chiefly in the district north-west of London.

According to Mansfield [1] the principal types of conglom-

---

[1] *Bull. Mus. Comp. Zool.*, *Harvard*, 49, Geol. Ser., viii, No. 4, 1906, pp. 105-151.   For petrology of conglomerates, see A. Hadding, *Lunds Univ. Aarskrift N. F. Avd.*, 2, Bd. 23, 1927, pp. 1-171.

erate are of marine, fluviatile, estuarine, lacustrine, or glacial origin. Of these, fluviatile conglomerates are by far the most abundant and important. True conglomerates composed of rounded fragments always indicate long-continued transport, mostly in water.

ARENACEOUS ROCKS. SANDS AND SANDSTONES—Sands [1] have already been treated to some extent in earlier sections (pp. 191-4). Sands may be classified as marine, estuarine, lacustrine, fluviatile, desert, fluvioglacial, and volcanic. The first four groups are closely allied in origin and characters, and might be classed together as aqueous sands. The grains are generally subangular and well-sorted (Fig. 54 A). Each type may, however, contain organic fragments consonant with the environment in which it was accumulated. Coastal dune sands, which are often classed with desert sands as æolian or wind-formed, are mainly aqueous sands drifted some little distance by wind action; and have suffered so little further rounding that in most cases they cannot be distinguished from marine sands. Desert sands show the most perfect rounding of grains, which is carried down to particles of very small dimensions (Fig. 54 B). They are frequently very free from dust or mica-flakes, these constituents having been blown away by the persistent desert winds. Fluvioglacial sands are distinguished by the sharp angularity of their fragments, and their unsorted character (Fig. 54 c). Most glacial sands yield a very large and varied suite of heavy minerals. Volcanic sands accumulate around volcanic islands, and are to be identified by the igneous nature of their constituents, their angularity, and good stratification. They pass imperceptibly into marine sands, fluviatile sands, etc.

Some sands are distinguished by the concentration of some one constituent, forming magnetite sand, garnet sand, monazite sand, etc.; and others by abnormal composition, as sands composed of organic fragments, or of basalt, etc.. A sand composed entirely of fragments of *Lithothamnion*, a cal-

[1] General accounts of sands are given by P. G. H. Boswell, "Sands : Considered Geologically and Industrially under War Conditions," Inaug. Lecture, Univ. Liverpool, 11 Nov., 1917 (1919), pp. 38 ; W. H. Scherzer, "Criteria for the Recognition of the Various Types of Sand Grains," *Bull. Geol. Soc. Amer.*, 21, 1910, pp. 625-62.

careous alga, forms part of the 10-foot raised beach in the Great Cumbrae island in the Firth of Clyde. In South Bute, the erosion of the Carboniferous basalt lavas has given rise to a marine sand composed of basalt particles.

The cementation of sand grains to form *sandstones* is alluded to in a previous section (p. 196).[1] Since sandstones may be formed by the cementation of any kind of sand, they may also be classified as marine, fluviatile, desert, etc. In addition to the common siliceous, calcareous, argillaceous, and ferruginous cements, other substances such as gypsum and barytes may occasionally act as binding materials. The cement is usually deposited between the grains, but silica may occasionally form outgrowths upon the quartz grains in optical continuity with them. When the interspaces are completely filled, a very solid, compact, hard rock is then formed, which is termed *quartzite*.

Ganister is a local name (Yorkshire) for certain very pure quartzites of the Carboniferous formation. Sandstones cemented with calcareous matter are very common, and are called calcareous sandstones. Whole formations, as, for example, the Calciferous Sandstone of the Lower Carboniferous of Scotland, and the Calcareous Grit of the Yorkshire Jurassic, are composed of this type. Ferruginous cements usually take the form of thin films of iron oxide around each grain, and produce red or brown sandstones, of which good examples are those of the English Triassic (Mansfield), and of the Permian in south-west Scotland (Locharbriggs, Ballochmyle). Argillaceous material does not often form a strong bond for sandstones, and when it is abundant the sandstone tends to crumble and break down into a sand. The so-called "rotten-rocks" of the Scottish Carboniferous are of this type, and are easily crushed down so as to make excellent moulding sands for steel manufacture.

A rather coarse sandstone with angular grains is known as a *grit*. Sandstone with abundance of felspar, usually derived from the disintegration of granite, is an *arkose*.[2] Much of the Torridonian sandstone of the north-west of Scotland is an arkose, having been derived from the breaking-up of the Lewisian granite-gneiss. *Graywacke* is sandstone, dark in

---

[1] For petrology of sandstones, see A. Hadding, *Lunds Univ. Aarskrift N. F. Avd.*, 2, Bd. 25, 1929, 287 pp.

[2] D. C. Barton, *Journ. Geol.*, 24, 1916, p. 417, has given a general account of arkose.

colour as the name indicates, derived from the waste of an area of miscellaneous rocks, including basic igneous types, slates, and sandstones. The basic rock chips and mineral fragments are collectively equal to, or in excess of, the quartz grains. Graywacke is an important lithological component of the Palæozoic, especially of the Lower Palæozoic, systems. Many of the red, purple, and brown sandstones of the Old Red Sandstone are also graywackes according to the above definition. Both arkose and graywacke consist of material which has undergone comparatively little transport, and the fragments are usually coarse, angular, and unsorted.

Many sandstones contain layers in which flakes of white mica are abundant (micaceous sandstone). The mica may be mixed with silty or clayey material in thin beds or laminæ, and may cause the rock to split into thin slabs which are coated with the spangles of mica (*faikes* in Scotland). A sandstone of this character with a calcareous cement, which splits readily along the micaceous layers into slabs suitable for paving, is called flaggy sandstone or *flagstone*.

A *freestone* is a uniform thick-bedded sandstone with few divisional planes. It can be cut or worked easily in any direction, and consequently forms a good building stone. The term freestone is also applied to some limestones of similar character.

SILT AND SILTSTONE—Although they form an important part of the sedimentary group, rocks of silt grade have hitherto not been accorded as much attention as those of sand or clay grade. Silts occur abundantly as the products of fluviatile, lacustrine, glacial, and æolian action. Many volcanic dusts are of silt grade, and should be included here. Silts are of finer grain than sandstones, but coarser than mud or clay. They often exhibit very perfect lamination and are earthy in texture, lacking the plasticity of clay, and the rough feel of sandstone. Some of the upper glacial brick-earths of East Anglia, especially the red and buff varieties, approximate to true silts, nearly 90 per cent. of the grains falling within the silt grade. They are regarded as due to the washing of boulder clay by glacier water in front of the Pleistocene ice sheet.[1]

[1] P. G. H. Boswell, " Petrology of the North Sea Drift, ' *Proc. Geol. Assoc.*, 27, pt. 2, 1916, pp. 88-9.

As an example of wind-formed silt, the *loess*, a fine cal-careous sediment of yellowish colour, which forms thick, uniform, unstratified sheets in Central Europe, Asia, and the United States, may be cited. It is regarded as due to the re-distribution by wind of fine sediments deposited in the glacial period, during a succeeding warm, dry, interglacial episode.[1] Many lake deposits and river alluvia will be found to be mainly of silt grade when their mechanical composition comes to be adequately investigated.

Compacted silts are quite common in the sediments of past geological ages; and for them the appropriate term *siltstone* may be employed. These rocks are widespread in many Palæozoic formations, where they have been described as gritty slates, fine-grained sandstones, gray-wackes, quartzites, slaty graywacke, etc. In Prince Charles Foreland (Spitsbergen) rocks of this type are char-acteristically laminated, and are interbedded with mudstones and sandstones, producing well-banded rocks.[2] The most abundant sedimentary types in South Georgia are fine, slaty grits and graywackes which should be described as siltstones.[3]

The Triassic marls of the British Isles appear to be largely loess-like sediments due to the trapping of wind-blown dust in wide, shallow, desert lakes like those of Western Australia. Professor R. A. Daly has described dense, compact quartz-ites of the Beltian system (Pre-Cambrian) in the Rocky Mountains region of Canada, which consist of angular quartz and felspar grains averaging ·05 to ·1 mm. in diameter. The homogeneity, thick bedding, mineralogical characters, and grade of these rocks, suggest that they may represent deposits of loess-like origin.[4] Finally, it is probable that many of the dark, slaty flagstones, with abundant lime and organic matter, of the Orcadian Old Red Sandstone, represent silty river alluvia of that period.[5]

[1] F. V. Emerson, "Loess-depositing Winds in Louisiana," *Journ. Geol.*, 26, 1918, pp. 532-41.
[2] G. W. Tyrrell, "Geology of Prince Charles Foreland," *Trans. Roy. Soc. Edin.*, 53, pt. 2, 1924, p. 461.
[3] G. W. Tyrrell, "Petrology of South Georgia," *ibid*, 50, pt. 4, 1915, p. 824.
[4] *Geol. Surv. Canada*, *Mem.* 68, 1915, pp. 102-6.
[5] *Mem. Geol. Surv. Scotland, Geol. of Caithness*, 1914, pp. 89-97.

ARGILLACEOUS ROCKS.   CLAYS AND SHALES—In this group
are included all detrital deposits and their compacted representatives, in which the average size of grain is less than
·01 mm.   When loose and dry this material forms *dust ;*
with varying amounts of moisture it forms *mud* and *clay*.
When welded into a compact rock argillaceous material is
called *shale* when it is well bedded and splits easily along the
bedding planes, and *mudstone* when it lacks the fissility of
shale, although it may or may not be well stratified.

The minerals of argillaceous rocks are frequently difficult
to identify because of their extremely fine state of subdivision.   Generally speaking, they consist of two main
groups ; the hydrated silicates of aluminium, hydrated iron
oxides, etc., produced by the decomposition of rocks, and the
" rock-flour " of comparatively fresh mineral fragments produced by disintegration.   Some calcareous and carbonaceous
matter, along with finely-divided sulphide of iron, are to be
found in many clays.   Argillaceous rocks are, therefore,
composed of the very finest particles liberated by rock
weathering.   Their chemical and mineral composition may
vary enormously according to their parentage, and according
to the relative proportions of the two main groups of constituents.   A common chemical character is a relatively high
proportion of alumina, resulting from the concentration of
hydrated aluminium silicates and finely-divided mica.
Furthermore, clays often show a preponderance of potash
over soda, owing to the fact that potassic salts are more
readily adsorbed by the colloidal matter in clays than are
the sodic salts.

Most clays, but not all, possess the characteristic property
of *plasticity* when wet.   Kaolin, for example, the purest type
of clay, is not plastic.   On the other hand, the slime produced
by the fine crushing of gold quartz is very plastic.   The
cause of plasticity is not yet well understood.   Apparently
it depends neither on composition nor on grain size.   It
probably has some connection with the presence of colloidal
matter in clays, and with its property of retaining moisture
and adsorbable substances.

Dusts, muds, and clays may be classed according to their
modes of origin, as marine, fluviatile, lacustrine, glacial,
æolian, or volcanic.   Dusts are formed on dry land, hence only

the agency of wind or volcanoes can be invoked for their production. Mention has already been made of dust storms in the United States, which carry millions of tons of material (p. 177). Some loess deposits, which are dust-like in fineness, also fall into the æolian category. The finest products of volcanic explosion rarely accumulate by themselves, but they contribute to other sediments, and form a considerable part of those formed in the abyssal depths of the oceans.

Clays of glacial origin almost always require the co-operation of water for their deposition. The rivers discharged from glacier fronts are laden with great quantities of fine rock flour which gives the water a milky appearance. The particles settle when the velocity abates sufficiently, and this occurs as the river divides into distributaries, or as it enters a lake. Hence glacially-produced silts and muds form large contributions to lacustrine deposits of like grade.

The *boulder clay*, however, is the most typical argillaceous rock of glacial origin. Material of clay grade, while usually forming the bulk of the deposit, is mingled with material of all other grades, especially pebbles, cobbles, and boulders of all sizes, scattered pell-mell through the fine-grained matrix. This description applies only to low level boulder clay, the product of the great ice-sheets in regions of low relief. Traced to the mountains the boulder clay, so-called, loses much of its clay base, and becomes a loam or even sand, through which boulders are scattered. This material is better called *glacial drift*.

Boulder clay is frequently associated with finely-laminated, fluvioglacial sediments, with alternating bands usually of silt or clay, each pair of which is supposed to represent the seasonal deposits of a year. By the study of these *varve* sediments G. de Geer[1] has been enabled to estimate the length of time taken by the retreat of the Pleistocene ice sheet in Scandinavia.

Muds and clays of fluviatile origin, with silts, form the *alluvium* found along the lower courses, and on the flood-plains, of large rivers. They vary greatly in composition, and usually carry a large quantity of organic, and especially vegetable, matter. Many delta and lacustrine deposits are

[1] *Compt. Rend. Congr. Internat. Géol.*, Sess. II, 1910, pp. 241-53.

of essentially similar character. Lakes act as settling tanks for rivers, and lacustrine silts and muds, which are spread slowly and evenly over the whole floor of the lake, have a tendency to very fine thin lamination.

The sea, however, is the great repository for argillaceous sediments. Marine muds are chiefly deposited between the 100-fathom line (the mud line) and the 2500-fathom contour line. At greater depths only the abyssal oozes are found (p. 236). Murray and Renard recognise five types of marine muds : [1] blue mud, red mud, green mud, coral mud, and volcanic mud. Of these, the first-named is by far the most abundant and important. It owes its colour to finely-divided organic matter and sulphide of iron. Red mud is confined to limited areas off the mouths of great rivers, such as the Amazon and Hoang-Ho. The colour is due to the presence of a considerable proportion of ferric oxide. Green mud owes its colour to the presence of the mineral glauconite (hydrated silicate of aluminium, iron, and potassium). It is usually more calcareous in composition than other marine muds. Coral muds and volcanic muds are formed in the oceanic areas around coral islands and volcanic islands respectively.

Clays, shales, and *mudstones* are consolidated dusts and muds. Clays may retain as much as 15 per cent. of water, but in shale and mudstones the process of welding into a firm rock has driven out much of the original moisture. As these rocks are derived from dust and mud, they may in the same way be classed as marine, fluviatile, etc. Their constituents are the same as in muds and dusts, with the addition of certain authigenous minerals. The process of induration seems to develop finely-divided impure white mica, chloritic substances, and also minute needles of rutile (TiO_2).[2]

The black and blue shales which are so common in the stratigraphical record are derived from the dark marine and estuarine muds rich in carbonaceous matter and sulphide of iron, similar to the muds now forming in the Black Sea, and

---

[1] " Rept. on Sci. Results of the Voyage of H.M.S. ' Challenger '. Deep Sea Deposits," 1891, p. 186.

[2] A. Brammall, " Reconstitution Processes in Shales, Slates, and Phyllites," *Min. Mag.*, xix, 1921, p. 212.

in swamp waters on the Esthonian coast of the Baltic.[1] The oil shales are still richer in organic matter, and on distillation yield oil and ammonium sulphate. Mr. E. H. Cunningham Craig regards oil shales as fossil oil rocks (rocks impregnated with petroleum) in which inspissated oil has been retained by adsorption in colloidal clay matter.[2] From microscopic study, however, Mr. H. R. J. Conacher [3] has been able to show that the oil-yielding materials in oil shale are resinous fragments derived from vegetable matter entombed in the mud. He believes that the Scottish oil shales originated as estuarine mud-flats to which an abundant supply of finely-macerated vegetable material was contributed by waters flowing from swamp areas.

Tillite is hardened boulder clay, which has now been found at many horizons in the geological record. It is frequently accompanied by banded mudstone representing finely-laminated varve sediments.[4]

China clay or *kaolin* is a white non-plastic material consisting, in the raw state, of the mineral kaolinite (pure hydrated silicate of alumina, $Al_2O_3,2SiO_2,2H_2O$), mingled with fragments of quartz, felspar, mica, etc., the residual minerals of granite. It is believed to be due to the action of hot vapours containing boric and hydrofluoric acids (pneumatolysis, p. 324), on the felspars of granites, and is associated with other pneumatolytic minerals such as tourmaline, topaz, fluor-spar, and tinstone.

Pottery clays are very plastic and highly aluminous clays which are practically free from iron. *Brick clays* are the more abundant varieties of clay containing fluxes, especially compounds of iron and magnesium, which promote fritting or incipient marginal fusion of the particles when they are burnt, thus binding the particles firmly together.

Fireclays are so-called because, owing to a low content of alkalies and other fluxes, they can be exposed to high temperatures without melting. They are used for making

---

[1] W. H. Twenhofel, *Amer Journ. Sci.*, 40, 1915, p. 272.
[2] *Proc. Roy. Soc. Edin.*, 36, 1916, p. 44.
[3] *Trans. Geol. Soc. Glas.*, 16, pt. 2, 1917, pp. 164-92.
[4] R. W. Sayles, " Seasonal Deposition in Aqueo-glacial Sediments," *Mem. Mus. Comp. Zool., Harvard*, 47, 1919, 67 pp.; T. W. E. David, " The Varve Shales of Australia," *Amer. Journ. Sci.*, 3, 1922, p. 115.

certain kinds of hard refractory bricks for furnace linings, sewer pipes, chimneys, and pipes which carry off chemically-active gases. For these purposes the clay must contain practically negligible quantities of alkalies, iron, and magnesia, and free silica must be present only in small amount. The latter requirement is very important, and great technical skill is needed to mix the various natural clays so as to get the right proportions of free silica. In England fireclays often occur immediately beneath coals, and probably represent the soils out of which grew the Carboniferous forests. In Scotland, however, the important Lanarkshire fireclays have no connection with coals, and are believed to have been deposited in lagoons.[1]

Marl is a clay rock which contains a considerable proportion of carbonates of lime and magnesia. The Triassic marls of England have already been mentioned (p. 211). *Marlstone* denotes an indurated marl. By an increase in the proportions of calcareous matter, marlstone passes over to argillaceous limestone, of which examples are to be found in the Lias, and in the Carboniferous of Scotland, where they are called *cement-stones*, because the clay and lime are frequently present in the right proportions for the manufacture of cement. Some cement-stones and marls are dolomitic, as, for example, those of Ballagan in Scotland.[2]

---

[1] J. W. Gregory, "The Glenboig Fireclay," *Proc. Roy. Soc. Edin.*, 30, pt. 4, 1910, pp. 348-60.
[2] *Mem. Geol. Surv. Scotland*, "Geology of the Glasgow District," 1925, pp. 11-12.

CHAPTER XIII

DEPOSITS OF CHEMICAL ORIGIN

CHEMICAL DEPOSITS IN GENERAL—The rocks treated in this chapter represent those products of rock weathering which are carried away in aqueous solution, and deposited directly by physico-chemical processes such as evaporation and precipitation. As precipitation from solution generally produces a finely divided crystalline or amorphous powder, the grain-size of the rocks formed by this process is usually small. On the other hand, the evaporation of solutions may, under favourable circumstances, promote the growth of large crystals, as in many salt and gypsum deposits. The recrystallisation of fine-grained chemical deposits favours the same development, and oolitic and pisolitic structures also produce an effect of coarse grain. Grabau has proposed the terms *spheryte*, *granulyte*, and *pulveryte* for chemical (and organic) rocks, the grain of which corresponds with that of rudaceous, arenaceous, and argillaceous rocks respectively.[1]

Evaporation and precipitation usually occur under conditions unfavourable to the development of well-shaped crystals. The growth of each crystal is hindered by the juxtaposition of its neighbours, and the resulting shapes are irregular. Precipitation often furnishes material so fine that the shape of the grains ceases to be a factor in the texture of the rock. Crystallisation may start from a centre and proceed radially, as in some stalactites or concretions; or it may take place intermittently around a nucleus, producing concentric arrangements of crystals (oolites, pisolites); or there may result a layered rock owing to interrupted deposition upon a surface or within a cavity (stalagmite, onyx, agate).

[1] *Principles of Stratigraphy*, 1913. p. 283.

CONCRETIONS [1]—Concretions are masses of differing chemical and mineralogical nature to the enclosing rock. They are generally rounded or nodular, but occasionally show fantastic shapes, which have figured in the folklore of primitive peoples (fairy-stones; *loess-puppchen* or loess-dolls). Concretions may vary in size from diameters of less than 1 inch to many feet. Their structure is often concentric around some nucleus, a fossil or inorganic fragment. Chemically, they generally represent a concentration of one of the minor constituents of the enclosing rock. In limestone and chalk, for example, there are nodules of chert or flint; in clays the concretions are calcareous or consist of sulphide of iron (marcasite); in the Carboniferous and Jurassic shales sphærosiderite (carbonate of iron) concretions are found; and in sandstone the concretions are occasionally of iron oxide, but more often carbonate of lime (whinny-boles). These substances are collected by percolating waters from the surrounding rock, and redeposited about some precipitating nucleus. Occasionally the concretionary material has been introduced from outside.

Concretions are usually externally smooth and compact, but often they are internally fissured, and the cracks healed again with another mineral, which is frequently calcite. Such are the septarian nodules or septaria of the London Clay.

SECRETIONS—Secretions are formed of substances deposited within empty cavities in any rock. The space may be completely or only partially filled with mineral matter carried in by percolating solutions, which is often deposited in bands parallel to the walls of the cavities. Amygdales and agates are secretionary structures found in igneous rocks. A *druse* is a cavity lined with minerals identical with those of the enclosing rock. Sometimes these hollow secretions are readily separable, and are then called *geodes*. The *potato-stones* of the Carboniferous Limestone in Somersetshire are geodes consisting of a shell of hard, silicified limestone, with an internal lining of quartz crystals.

Dendrites are secretionary forms of iron and manganese oxides which are found as delicate plumose and arborescent

---

[1] On concretions see W. A. Richardson, *Geol. Mag.*, lviii, 1921, pp. 114-24.

growths resembling moss or branches. They coat the faces of fissures and parting planes, but occasionally penetrate the substance of a rock, especially in fine-grained, earthy limestones. In polished sections these dendrites simulate hedgerows, lines of trees, like a fen country landscape, producing the well-known *landscape marble.*

The coloration [1] (p. 221) and decolorisation of rocks are due to chemical phenomena of some complexity. The *bleach-spots,* or *deoxidation-spheres,* which are commonly seen in red rocks of all kinds, are probably due to the reduction of ferric oxide around fragments of organic matter, and the subsequent removal of the iron in the more soluble ferrous form by percolating waters. A very fine example in sandstone, showing the Liesegang banding (p. 222), is illustrated in Fig. 62.

The classification of deposits of chemical origin naturally rests upon a chemical basis. The one adopted here is as follows :—

1. Siliceous deposits.
2. Carbonate deposits.
3. Ferruginous deposits.
4. Salts.
    (a) Chlorides, sulphates, and carbonates.
    (b) Borates.
    (c) Nitrates.

COLLOIDS [2] (Gr. *kolla* = glue)—Heterogeneous substances consisting of two phases, one of which is dispersed as discrete particles through a continuum of the other, are called *dispersed systems.* These may be coarse, colloid, or molecular, in decreasing order of size. A magma containing crystals may be regarded as a coarse dispersed system. When the sizes of the particles dwindle down to ultramicroscopic dimensions, that is, to diameters between $1/1,000,000$ mm. and $1/10,000$ mm., a colloidal system is produced. The particles in a colloid cannot be caught on an ordinary filter, but they

[1] F. H. Lahee, *Field Geology,* 2nd ed., 1923, pp. 18-20 ; W. A. Richardson, " Petrography of the Marlstone Ironstone of Leicestershire," *Trans. Inst. Min. Eng.,* lx, pt. 4, 21, pp. 337-44.

[2] E. Hatschek, *Introduction to the Physics and Chemistry of Colloids,* 1918 ; A. Holmes, *Nomenclature of Petrology,* 1920, pp. 63-4 ; P. Niggli, *Lehrbuch d. Min.,* 1921, pp. 430-45 ; D. Hubbard, "Colloids in Geologic Problems," *Amer. Journ. Sci.,* 4, 1922, pp. 95-110 ; A. Scott, "The Application of Colloid Chemistry to Mineralogy and Petrology," *Rept. Brit. Assoc. Adv. Sci.,* 1922, pp. 204-43.

will not dialyse or diffuse through certain membranes such as egg-skin, gold-beater's skin, or parchment paper. Glue and gelatine are typical colloidal substances. The particles in colloids are believed to be aggregates of about 100 molecules. With still further decrease in size, molecular dispersions are arrived at, in which the phases are in true

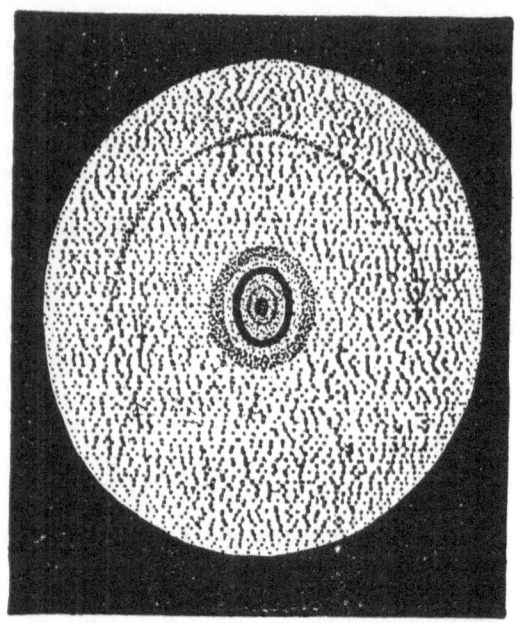

FIG. 62.—DEOXIDISATION SPHERE IN SANDSTONE.
Shows concentric banding. Permian red sandstone, Ballochmyle, Ayrshire. Diameter, 4¼ inches. (See text, p. 219.)

solution. Coarse dispersions are distinguishable from colloids by the fact that their particles are microscopically visible. Molecular dispersions or solutions are distinguished from colloids by their ability to pass through membranes. Liquid colloids are called *sols*; and when the particles of a sol coagulate or settle down to a gelatinous mass, the product is known as a *gel*.

The table [1] below gives some examples of various types of dispersed systems.

$$\left.\begin{array}{l} s = \text{solid} \\ l = \text{liquid} \\ g = \text{gas} \end{array}\right\} \text{dispersed in a medium of} \left\{\begin{array}{l} S = \text{solid.} \\ L = \text{liquid.} \\ G = \text{gas.} \end{array}\right.$$

System.	Coarse.	Colloid.	Molecular.
s in S	Solid inclusions in minerals.	Blue rock salt (metallic Na in NaCl).	Solid solutions.
	Suspensions :		
s in L	Magma with crystals.	Suspensoids.	Solutions.
s in G	Dust ; volcanic ash.	Smoke.	—
l in S	Liquid inclusions in minerals.	Occluded liquids.	Water of crystallisation.
l in L	Emulsions.	Emulsoids	Solutions.
l in G	Rain.	Fog.	—
g in S	Gas inclusions in minerals.	Occluded gases.	Adsorbed gases.
g in L	Foams.	Colloid foams.	Solutions.
g in G	—	—	Solutions.

Many minerals and rocks are now known to be of colloidal origin, and colloids enter into many geological processes. During the cutting of the Simplon tunnel a vein of gelatinous silica was actually found, and it is highly probable that similar material was the source of such minerals as opal, agate, chalcedony, and some quartz.

The sols and gels occurring in Nature have an aqueous medium, and are therefore designated *hydrosols* and *hydrogels*. A phenomenon of considerable geological importance is the ease with which solutions, especially of salts, diffuse through hydrogels. If two different solutions are diffusing through a gel precipitation may take place, and it is suggested that the metallic gold and copper found in some quartz veins is due to the reduction of their solutions in a silica gel. The fine blue colour of some halite (rock salt) has been shown to be due to metallic sodium in the colloidal state. Similarly, colloidal particles of carbon and iron oxide cause the smoky

[1] From Holmes, *op. cit. supra.*

colour of cairngorm, and colloidal carbon produces the colour of amethyst.   The red, brown, and yellow colourings of rocks are due to films of colloidal ferric oxide and hydroxide.

Certain banded structures in minerals and rocks are probably due to rhythmic precipitation in gels.   The reactions which produce the banding no doubt take place as in the phenomenon known as Liesegang's rings.[1]   These are produced by the diffusion of a solution through a gel which already contains a substance with which the solution can react to form a precipitate.   As the solution advances the new substance produced by the reaction at first forms a supersaturated solution, which reaches precipitation at a certain stage. This uses up all the material in the neighbourhood, and precipitation ceases until the diffusing solution has advanced far enough to produce another zone of precipitation. Banding is thus caused, the features of which are conditioned by all the circumstances of diffusion, as for example, the point or line of entry of the solution, and the shape of the vein or cavity in which the reactions take place.   The rusty bands which are often seen to be roughly parallel to the exterior of a weathering block, are probably due to rhythmic precipitation of colloidal iron oxide (Fig. 63).

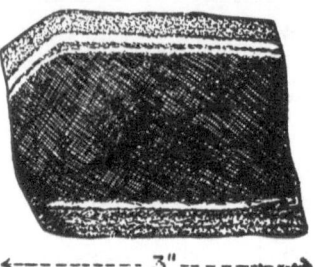

<----------- 3" ------------>

FIG. 63.—RHYTHMIC COLOUR BANDS DUE TO WEATHERING.

Rhythmic precipitation bands on the weathered edges of a block of riebeckite-orthophyre, Holy Isle, Arran.   The fresh rock is dark blue-grey in colour.   Weathering causes pinkish-brown, purplish-brown, and greyish-white colour-banding parallel to the weathered faces.

In time hydrogels lose a portion of their water and harden, producing various types of amorphous minerals (opal, limonite).   As colloidal systems are fundamentally unstable, further change leads to crystallisation, which may be fibrous

---

[1] Liesegang, *Geologische Diffusionen*, 1913.

at first, but may finally be coarse and granular. The two series Opal (amorphous) → Chalcedony (fibrous) → Quartz (crystalline), and Limonite (amorphous) → Kidney ore (fibrous) → Hæmatite (crystalline), are typical of this mode of change. Concretionary, botryoidal (grape-like), nodular, stalactitic, oolitic, and pisolitic structures, are characteristic of minerals and rocks of colloidal origin. The functions of colloidal matter in soils (p. 185) and clays (p. 212) have already been dealt with. For further discussion of colloids in relation to geological problems see the works cited in this section, and the succeeding sections of this chapter.

SILICEOUS DEPOSITS—While quartz is practically insoluble in water, the other forms of silica, chalcedony and opal, are fairly soluble, especially in the presence of alkaline carbonates. In some volcanic regions (Iceland, Yellowstone Park, New Zealand) hot springs bring up silica, and deposit it in mounds and terraces about the orifices of eruption. This material, which consists of cryptocrystalline or opaline silica, is called *siliceous sinter*. Its deposition is due mainly to evaporation and cooling of the waters ; but certain lowly plants, algæ, some species of which live in hot spring water, are also instrumental in precipitating the silica.

The most abundant siliceous deposits of chemical origin are *flint* and *chert*, which consist of minutely-crystalline or cryptocrystalline silica, and occur as irregularly-shaped nodules or tabular masses, whch are disposed in bands through limestone. Chert may also occur as a general replacement of limestone, and in Spitsbergen whole formations, up to 800 feet in thickness, are composed of chert. The term flint is usually restricted to the siliceous concretions of the Chalk, while chert is used for similar material occurring in other formations. The nodules of flint in the Chalk frequently form around fossils, such as sponges or echinoderms.

Many theories have been put forward to explain the origin of flint and chert, and these explanations fall into two main groups. In one, the siliceous material is ascribed to the original deposition of colloidal silica upon the sea floor, contemporaneously, or almost contemporaneously, with the associated limestone, and with some subsequent replacement of the limestone by the silica. The other theory places the deposition of the silica at a time long after the complete solidification and uplift of the limestone. The material is carried by percolating waters which collect the silica scattered through the rock, and re-deposit it in bands which recall the rhythm of the Liesegang rings.

The origin of the Chalk flint has recently been investigated by Dr. W. A. Richardson,[1] who has supported the view first advanced by Liesegang,[2] and extended by Cole,[3] that flint is due to the rhythmic precipitation of a silica solution diffusing through the Chalk. The amount of silica present as flint is found to be of the same order as that disseminated in other ways through the Chalk. Furthermore, there is a distinct inverse relation between the amount of disseminated silica and that segregated as flint. Hence the silica could easily have been derived from the siliceous organisms entombed in the Chalk, and only a moderate thickness of the formation would be necessary to supply the amount needed for flint. The flint occurs in lines of nodules which recur rhythmically. Starting from the highest flint band, the distance between adjacent bands decreases rapidly to a minimum which persists through a considerable thickness of the Upper and Middle Chalk; but towards the base of the Middle Chalk the distance between the bands widens out again. This rhythm is very like that obtained by the diffusion of solution in gels, and the suggestion is therefore put forward that flint represents rhythmic precipitation of silica from downward draining solutions produced during the uplift of the Chalk. In the early stages of the process there would be increasing concentration of the solutions concomitant with decreasing spatial separation of the flint bands. In later stages progressive dilution of the solutions would lead to a gradual lengthening of the intervals between precipitation.

Much chert is unquestionably due to the replacement of limestone by silica at a date long subsequent to the origin of the formation. Features indicative of replacement are the occurrence of chert along fissures; the irregular shapes of some nodules; the presence of patches of limestone in some chert masses; the association of chert with silicified fossils;

[1] *Geol. Mag.*, 1919, pp. 535-47.
[2] *Geologische Diffusionen*, 1913, p. 126.
[3] *Geol. Mag.*, 1917, pp. 64-8.

preservation of original structures and textures in chert; failure of some chert masses to follow definite horizons; and the occurrence of silicified oolitic limestone. Dean has shown that the growth of chert nodules in some Missouri limestones has not only arched up the overlying stratum, but has caused small faults.[1] This author has also shown by experiment that silica hydrosols are remarkably stable in the presence of calcite; but if they become saturated with carbon dioxide under the same conditions, the silica is quickly precipitated.

The theory of the primary colloidal deposition of chert at approximately the same time as the enclosing limestone has been put forward in America by W. A. Tarr,[2] and in Great Britain by H. C. Sargent. In regard to the Carboniferous cherts of Derbyshire, the latter author shows that some occur in the neighbourhood of former volcanic centres, suggesting that the silica emanated from submarine volcanoes and is therefore of magmatic origin.[3] The radiolarian cherts which are found associated with the Ordovician spilitic lavas of Ayrshire and other localities have also been ascribed to the same source. Siliceous emanations have provided the material for enormous swarms of radiolaria in the neighbourhood of submarine eruptions.[4] Sargent has also studied the massive chert formation in the Carboniferous of North Flintshire. This is regarded as an original chemical sediment of inorganic origin laid down in a nearly landlocked basin.[5]

CARBONATE DEPOSITS—In this section are treated deposits of calcium and magnesium carbonates. Iron carbonate, however, is dealt with under the heading of ferruginous deposits, and sodium carbonate under salts. For the deposition of calcium carbonate organic agencies are chiefly responsible (p. 235), yet purely inorganic deposition brought about by changes in physico-chemical conditions may occur pari passu with, or quite independently of, organic formation. It has been shown that in the ocean the surface layers of

[1] Amer. Journ. Sci., 45, 1918, pp. 411-5.
[2] Ibid., 44, 1917, p. 409.
[3] Geol. Mag., lviii, 1921, pp. 265-78.
[4] Dewey and Flett, "British Pillow Lavas," Geol. Mag., 1911, p. 245.
[5] Geol. Mag., lx, 1923, pp. 168-83.

15

water are substantially saturated with calcium carbonate, except in the polar regions and within cold currents. Under these conditions, loss of carbon dioxide, or a rise of temperature, or both acting together, result in the precipitation of this substance. Some fine-grained, non-fossiliferous limestones are certainly to be ascribed to this process; but it is difficult, if not impossible, to separate purely inorganic deposition from that due to biochemical processes (p. 234).[1]

In fresh waters the precipitation of calcium carbonate is chiefly due to the loss of carbon dioxide. Almost all natural waters contain this salt, and the amount that it is possible for the water to dissolve is greatly increased if carbon dioxide is also present. The loss of the gas by the evaporation of water dripping from the roofs of limestone caves, and flowing over their floors, leads to the formation of the well-known *stalactites* and *stalagmite*, by the gradual accretion of calcium carbonate separated as films over successive drops. Stalactites are the long, icicle-like pendants from the roof; stalagmite is the material formed on the floor of the cave. Stalagmite is often finely banded, and may take a good polish, when it is called *onyx*.

By the relief of pressure as it rises to the surface, spring water charged with calcium carbonate may reach saturation in that substance, which is then deposited around the orifice of the spring. In this way large deposits of porous cellular limestone are built up. This material is called *calc-tufa*, *calc-sinter*, or *travertine*. Calcareous algæ and bacteria have been found to be active in the formation of some deposits of calc-tufa. Rivers flowing through limestone country may also deposit calc-tufa in thick sheets which sometimes cause waterfalls.[2]

In tropical countries where a rainy season is followed by a long, dry season, ground-water saturated with calcium carbonate is brought to the surface by capillarity during the dessication period, and the carbonate is deposited just below the soil as a hard layer which is often nodular and rich in iron. This material is called *kunkar* in India. The *corn-*

---

[1] J. Johnston and E. D. Williamson, " Rôle of Inorganic Agencies in the Deposition of Calcium Carbonate," *Journ. Geol.*, 24, 1916, pp. 729-50.
[2] J. W. Gregory, " Constructive Waterfalls," *Scott. Geogr. Mag.*, 27, 1911, pp. 537-46.

stones of the Old Red Sandstone, and the impure calcareous concretions of certain beds in the Trias, are probably fossil representatives of kunkar.

Some limestones consist largely of minute spherical or ellipsoidal grains of calcium carbonate which resemble fish roe. Hence the grains are called *ooliths* and the rock containing them *oolite* or *oolitic limestone*. Ooliths generally show a series of concentric coats of calcareous material in which a radiating crystalline structure can often be made out, and they may or may not be built around a nucleus such as a mineral grain or a fragment of shell. Pisoliths are essentially similar to ooliths, but reach much larger sizes, and are generally found in residual deposits (p. 182). Ooliths are often broken and worn, and the limestones built of them show current bedding and other signs of deposition in shallow waters.

Ooliths are forming at the present day on the shores of lakes, such as the Great Salt Lake of Utah, where shallow water saturated with calcium carbonate is in a state of constant wave agitation. The carbonate is deposited on sand grains or shell fragments, and the rounded shapes of the grains are supposed to be due to their continual abrasion through movement. Linck [1] found that all recent ooliths he examined were composed of the aragonite form of calcium carbonate, whereas the older fossil occurrences were calcite. This fact was confirmed by direct experiment on the precipitation of calcium carbonate from sea water. Ooliths are consequently formed as aragonite, but change in course of time to the more stable form calcite.

More recently Bucher [2] has advanced the view that ooliths, like pisoliths, are to be regarded as due to colloidal precipitation, and are to be interpreted, as Schade has done for the analogous urinary calculi and concrements, as concretionary bodies which are due to the solidification of an emulsoid (p. 221). When a change towards the crystalline state occurs, a radial structure develops if the substance is pure; but if other substances have been precipitated along with it, a concentric structure is produced. The spherical shapes of

[1] *Neues Jahrb. f. Min. Beil.-Bd.*, 16, 1903, p. 495.
[2] *Journ. Geol.*, 26, 1918, pp. 593-609.

the grains are regarded as due to the tendency of the colloidal droplets to coalesce, and exhibit surface tension phenomena.

In another view of the origin of ooliths filamentous algæ are supposed to have been active. Some ooliths contain sinuous fibres or tubes like those of the *Girvanella* type of calcareous algæ. There is no doubt that calcareous algæ abound in waters from which ooliths are precipitated, but the modern view is that their enclosure within the ooliths is merely accidental, and that the incrustation of minute ooliths or organic fragments by algæ is the explanation of certain types of supposed oolitic grains.[1]

The term *dolomite* should be restricted to the mineral species the composition of which is the double carbonate of calcium and magnesium ($CaCO_3,MgCO_3$). It has also been extended to the massive rock consisting essentially of dolomite, but the term *dolomite-rock* should be employed in this sense. *Dolomitic limestone* is a limestone in which part of the calcite has been replaced by dolomite ; and *magnesian limestone* should be restricted to limestones which contain a notable amount of magnesium carbonate, although dolomite itself is not present. The majority of these rocks originate through the replacement of ordinary limestone by magnesium carbonate, which is brought in by solution in ground waters, or in the case of marine types, is obtained from the sea water. In the first case the limestone is changed along joints and fissures. With complete replacement a volume contraction of 12·3 per cent. must ensue, resulting in the production of an open porous rock. In the sea coral-reef limestones are especially liable to *dolomitisation*. Alcyonarian corals contain from 6 to 16 per cent. of magnesium carbonate to start with, and as calcite is more soluble than magnesium carbonate a progressive enrichment in the latter may be presumed to take place by leaching. Beside this, there is direct replacement of the calcium carbonate by magnesium carbonate, especially between the surface and 150 feet of depth ; and the reaction is facilitated by the fact that the corals at first consist of the unstable form aragonite.

Whether dolomitic rocks can be formed by direct chemical

[1] *Journ. Geol.* 26, 1918, p. 606; also see Harker, *Petrology for Students*, 3rd ed., 1902, Fig. 60, p. 257.

precipitation is a moot question. Fresh water dolomite rock
has been described, and Professor C. G. Cullis has shown that
the Trias Marl of the English Midlands contains numerous
minute, perfect rhombs of dolomite.[1]   The German Trias
contains beds of dolomite rock associated with rock salt,
gypsum, and anhydrite, suggesting direct precipitation of
the rock in highly-saline relic seas.

FERRUGINOUS DEPOSITS—Iron salts are present in most
natural waters, and under suitable conditions ferruginous
materials are deposited as oxides, hydroxides, carbonates,
or silicates.   The iron is in solution mainly as bicarbonate,
less commonly as chloride or sulphate.   Loss of carbon
dioxide converts the bicarbonate into ferrous carbonate, and
as this salt oxidises almost instantly in contact with air the
waters of springs, streams, or lakes containing this salt deposit
rusty films of iron hydroxide, which collect along the banks
of streams or on the floors of lakes, yielding a soft, porous
material known as *bog iron ore*.   If precipitation occurs under
reducing conditions, the ferrous carbonate itself may be
deposited.   The *clay ironstone* and *blackband ironstone* of the
Coal Measures consist of siderite granules intermingled with
clayey and coaly matter.   They are believed to have been
deposited under swamp and lagoon conditions from iron-
bearing solutions by loss of carbon dioxide coupled with a
reducing environment.[2]

In marshy and peaty soils, as well as in the alluvium of
semi-arid regions, it is common to find a hard layer of iron
oxides a foot or so beneath the surface.   This *iron pan* is
formed in exactly the same way as the calcareous kunkar
(p. 226), by the oxidation of ferriferous solutions drawn up
by capillarity.

Iron deposited as silicate generally takes the oolitic form
and is of marine origin.   Layers of chamosite (a mineral of
the chlorite family with the approximate formula, $2SiO_2$,
Al_2O_3, $3FeO$, aq.) are believed to form around nuclei on the
sea floor, or as concretions in soft ooze, under conditions in

---

[1] *Rept. Brit. Assoc. for Adv. of Sci.*, Leicester, 1907, p. 506.
[2] A. F. Hallimond, " Iron Ores : Bedded Ores of England and
Wales. Petrography and Chemistry," *Spec. Repts. Min. Res. Gt.
Britain*, 29, 1925. This work contains one of the best recent discussions
of iron-ore deposition.

which the sea water is saturated with ferrous salts and aluminosilicic acid, the latter probably in the form of colloidal clay matter.[1] Chamosite ooliths embedded in chamosite mud with detrital material and carbonates, build up great iron formations such as the famous Cleveland ironstone of Yorkshire, and the Wabana iron ore of Newfoundland.[2] Formerly these rocks were thought to be due to the replacement of oolitic limestone by iron-bearing solutions, but the primary character of the deposition is now fully recognised. They have been deposited as a loose, sea-floor aggregate of ooliths, shell fragments, rolled pieces of consolidated ironstone and other detritus; and the ferrous material was a chemical precipitate from sea water. The process of submarine rock weathering through which the iron has been concentrated in these sediments has been called *halmyrolysis*.[3]

It is probable that the chemical precipitation of iron salts is partly conditioned by the presence of algæ and bacteria, especially the iron bacteria, with which, no doubt, the theatres of deposition often teemed. These organisms have the power of secreting limonite from iron solutions. Exceptionally, bacteria are preserved in ironstones, and shell fragments in ironstones are riddled with tubes due to perforating algæ.

SALTS [4]—By salts is here meant the chlorides, sulphates, carbonates, nitrates, and borates of sodium and potassium, and the sulphates of calcium and magnesium. Large and important economic deposits of chlorides and sulphates are formed by the evaporation of waters of enclosed arms of the sea and of salt lakes. The average salinity of normal sea water is 35 parts per 1000, and the percentage composition of the salts is 77·76 $NaCl$, 10·88 $MgCl_2$, 4·74 $MgSO_4$, 3·60 $CaSO_4$, 2·46 K_2SO_4, ·34 $CaCO_3$, and ·22 $MgBr_2$. Before these salts can be deposited the water must become saturated with them, and as sea water is not nearly saturated, special geological conditions must supervene before their deposition

[1] Hallimond, *op. cit. supra.*
[2] A. O. Hayes, *Mem. Geol. Surv. Canada*, 78, 1915.
[3] K. Hummel, " Die Enstehung eisenreicher Gesteine durch Halmyrolyse," *Geol. Rundschau*, 13, 1922, pp. 41-81 ; 97-136.
[4] A. W. Grabau, *Geology of the Non-Metalliferous Mineral Deposits other than Silicates*, vol. i ; *Principles of Salt Deposition*, 1920, pp. 435.

takes place. An arm of the sea may be cut off by earth movements, or by the shallowing of its connection with the main body ; and if the climate is suitable the salts may be concentrated sufficiently for precipitation to occur. The general order of deposition is first lime salts, then sodium salts, magnesium salts, and finally, on complete evaporation, the potassium salts.

An example of the *modus operandi* of salt deposition is provided by the Karabugas Gulf on the east side of the Caspian Sea, from which the gulf is barred off by long, narrow spits broken through by a single channel. The gulf is surrounded on three sides by desert with a hot, dry climate. Hence there is active evaporation, and the water thus lost is supplied by a current from the Caspian which is estimated to carry in daily a load of 350,000 tons of sea salts. Thus a constant supply of salts is being carried in to the bay, but none is getting out, and the concentration is now such that gypsum ($CaSO_4$, $2H_2O$) is being deposited in considerable quantity, and sodium sulphate (mirabilite or Glauber's salt ; $Na_2SO_4, 10H_2O$) to the estimated amount of 1000 million metric tons.

The concentration of the Karabugas water is not yet sufficient for the deposition of sodium chloride (common salt), but in past geological times similar conditions have given rise to the great salt deposits of the Triassic in Cheshire, of the Permian in Germany, of the Miocene in Galicia, and the Silurian in North America. Practically the only great deposit of potash salts in the world is that of the Stassfürt region of Germany, which contains common salt, gypsum, anhydrite, with potash and magnesian compounds, as well as various iodides, bromides, and borates. It is probably due to the final complete evaporation, after several interruptions and regressions, of an extensive relic sea of Permian times.[1]

The deposits of terrestrial *salt lakes* are usually distinguished from marine salt deposits by the presence of sodium carbonate ; *bitter lakes* deposit sodium and magnesium sulphates in abundance ; in nearly all cases calcium sulphate

[1] F. W. Clarke, *Data of Geoch...mistry*, 5th ed., 1924, pp. 222-9 ; J. W. Gregory, *Trans. Geol. Soc. Glasgow*, 16, pt. 1, 1916, pp. 12-16.

(gypsum) is deposited. Terrestrial salts may be of *connate* origin, that is, derived from the salts included within the pore spaces of ancient marine sediments. This is unquestionably the origin of many of the very pure salt deposits of deserts, such as those of the Lop basin of Chinese Turkestan. The salt is leached from the rocks by the circulating ground water, but it can only be collected and concentrated in the inwardly-directed drainage basin of a desert ; or it may be drawn in solution to the surface by capillary action, leaving an efflorescence of salt upon evaporation.

Many salt deposits have been derived from the decomposition products of older rocks, which have been leached by circulating waters, and concentrated in lakes where they may reach saturation and thus be precipitated. The principal salts deposited in this way are the alkaline carbonates, nitrates, alums, chlorides, and sulphates. The soda and sulphate lakes of Western America and of Russia furnish illustrations of this mode of origin.

Of nitrates the best known and most extensive deposit is that found along the Pacific coast of Chili and Peru in the deserts of Atacama and Tarapaca. The main salt is sodium nitrate or *caliche* ($NaNO_3$). As this is an extremely soluble salt, it can only accumulate in practically rainless regions. The deposits are found on the sites of shallow lakes or playas, of which a characteristic section is as follows :—

Sand and gravel	1–2 inches.
Porous earthy gypsum	6 ,,
Compact earth and stones	2–10 feet.
Costra (low grade caliche with salt and detritus)	1–3 ,,
Caliche	1½–2 ,,
Clay	3 inches.

Common salt accompanies the nitrate in most cases, and forms deposits of much greater extent than the caliche.

The origin of these deposits is a question of considerable difficulty. Charles Darwin regarded them as of marine origin, and Noellner has ascribed them to the decomposition of great masses of sea-weeds which were left stranded by the recent emergence of land on the Pacific coasts of South America. The presence of iodine in the caliche deposits is in favour of this view.   Another theory ascribes them to the decomposition of guano deposits, but the absence of phos-phates renders this improbable.   W. Newton's explanation that the nitrates are derived from the oxidation of organic matter in the soil of the pampas, and are concentrated by the infrequent river floods which collect the nitrates from thousands of square miles, and deposit them in interior basins, is regarded by Clarke as the most feasible.[1]

Borates occur naturally in the form of *borax* ($Na_2B_4O_7$, $10H_2O$), and as the well-crystallised mineral *tincal* of the same composition.   They are deposited in lakes within regions of recent volcanic activity.   The fumeroles of Tuscany are jets of steam carrying boric acid, which is deposited in neigh-bouring lagoons.

A great variety of other natural salts have been described, but the majority are of purely local occurrence and of small petrological importance.

[1] *Data of Geochemistry*, 5th ed., 1924, pp. 254-9.

CHAPTER XIV

DEPOSITS OF ORGANIC ORIGIN

ORGANIC DEPOSITS IN GENERAL—In this chapter are treated those secondary deposits which are due, directly or indirectly, to the vital activities of animals and plants. These materials accumulate mainly on the sea floor, but fresh water and terrestrial examples are also well known. Only in rainless deserts and in the frozen polar lands are they absent. A rock of organic origin may be built up directly, from the beginning, as a quite solid material, as in the case of coral rocks and some algal limestones.[1] In other cases the deposition may be *biochemical* or *biomechanical ;* biochemical, when the vital activities of the organisms promote chemical conditions which favour precipitation, as in the cases of bacterial iron ores and limestones ; biomechanical, when the rock is due to the detrital accumulation of organic materials, as in the cases of crinoidal and shelly limestones, and some coals. It is obvious that, in this case, there may be gradual transitions to sedimentary types through an increasing admixture of inorganic detrital materials. In like manner biochemical deposits may pass into rocks of purely inorganic chemical origin. In many instances it is difficult to decide whether a given rock should be assigned to the sedimentary, chemical, or organic groups.

The grain-size of biomechanical rocks depends on the initial sizes of the component organisms, or of the fragments into which they naturally break. Thus foraminiferal, radiolarian, and diatomaceous deposits form oozes of clay grade ; while deposits of certain shells may be of extremely

[1] For this mode of origin Mr. H. D. F. Kitto, B.A., of the Greek Dept., Glasgow University, has kindly suggested the term *stereophytic* (= originating solid).

coarse grade (e.g. Hippurite limestone). The hard parts of other organisms, as, for example, those of crinoids, sea-urchins, etc., break up into comparatively large fragments, each consisting of a single crystal of calcite, and a coarse grain ensues in rocks resulting from their accumulation. Rocks of biochemical origin are usually fine-grained, as they are chiefly due to chemical precipitation (p. 217).

The classification of organic deposits is based upon their chemical composition, as follows :—

1. Calcareous . . Limestones.
2. Phosphatic . Phosphorites, Guano.
3. Ferruginous . Bacterial iron ores.
4. Siliceous . . Radiolarian and diatom oozes.
5. Carbonaceous . Coal, peat, etc.

A subsidiary grouping may be effected into *zoogenic*, deposits attributable to the agency of animals ; and *phyto-genic*, those due to the agency of plants.

ORGANIC ROCKS OF CALCAREOUS COMPOSITION : THE LIMESTONES—The organic limestones are composed chiefly of calcite ($CaCO_3$). Varying amounts of impurities may be present, giving rise to sandy, clayey, glauconitic, ferruginous, phosphatic, or bituminous varieties. Magnesia-bearing lime-stones may arise through the presence of organic material rich in magnesium carbonate (p. 228). In some limestones, as, for example, those built of corals, the calcium carbonate originally crystallised as aragonite, the orthorhombic form ; but as aragonite is unstable, and readily alters into the stable form calcite, it rarely survives in limestones. The organic limestones are mainly due to biomechanical processes; others are deposited from the first as solid rock (stereophytic) ; still others are formed by biochemical processes, and are very difficult to distinguish from those due to inorganic pre-cipitation.

Biomechanical limestones are frequently very hetero-geneous in composition, consisting of a great variety of organic fragments embedded in calcareous mud due to their comminution. One group of organisms occasionally pre-dominates, and gives character to the rock. The whole range of organisms which secrete hard parts are to be found in limestones. Foraminifera, corals, crinoids, mollusca, and crustacea are the animal classes most concerned in limestone formation; and the importance of calcareous algæ as limestone builders among the plants has recently been fully recognised. Other groups of calcareous organisms are locally abundant, and may be widely distributed in limestones, the main bulk of which is formed from the organisms mentioned above.

Limestones consisting largely or entirely of foraminifera are abundant. The *oozes* which cover great areas of the ocean floor are mainly calcareous and foraminiferal. The most widely spread of these deposits is the *Globigerina ooze*, which consists chiefly of the tests of Globigerina and other foraminifera forming part of the floating population (plankton) of the uppermost 200 yards of the ocean. This deposit is estimated to cover 50,000,000 square miles, mostly in the Atlantic, but also in the Indian and Pacific Oceans. It occurs at depths between 2500 and 4500 yards, but not at greater depths, as solution of the calcium carbonate of the shells increases rapidly with depth owing to the increase of pressure. Calcareous tests do not survive below 4500 yards. *Pteropod ooze*, consisting of the minute, delicate, molluscan shells called Pteropoda, may also be mentioned here. It is found at a depth of 1400 to 2700 yards, and is confined to the equatorial areas of the Atlantic Ocean. Oozes accumulate with extreme slowness. The trawls often bring up earbones of whales and shark's teeth which are lying on the surface of the ooze or are embedded in its uppermost layers; and these often show corrosion marks above the line up to which they were protected by the ooze. Raised fossil oozes at present above sea level are rare, but have been found in Barbados, Cuba, Borneo, and in some South Pacific islands.

Chalk may be regarded as a kind of consolidated ooze. The tests of foraminifera, especially Globigerina, are abundant, as well as shell fragments, sponge spicules, coccoliths, and rhabdoliths (p. 238), but the bulk of the deposit consists of finely-divided calcareous mud which may be due to the comminution of organic fragments. Foreign detrital matter is rare, but E. B. Bailey has pointed out the wide distribution of rounded and polished quartz grains in the Chalk, ranging from France to the Western Isles of Scotland, and has inferred that the Chalk sea was bounded by desert shores, and that the purity of the deposit was due to lack of rain resulting in the almost complete absence of terrigenous sediment.[1] Other foraminiferal limestones are the Nummulitic limestone of the Eocene, which covers vast areas around the Mediterranean, and the Saccamina and Fusulina limestones of the European Carboniferous.

Biomechanical *coral limestones* are due to the consolidation of coral detritus, which may be mingled with varying quantities of other organic fragments and mineral grains. They are produced by the pounding of surf against coral reefs, and the distribution of the resulting calcareous sand and mud by waves and currents as stratified deposits. The animals which live on and around coral reefs, especially the crustaceans, are known to feed upon the growing corals, and may produce in this way a large amount of detrital material.

Crinoidal limestones are due to the breaking-up of crinoid groves, examples of which exist in some tropical waters at the present day. The ossicles of the jointed stems, and the plates of the bodies, each consisting of a single crystal of calcite, form the main constituents of these limestones, which may, however, be mingled with the remains of other calcareous organisms, in a matrix of precipitated calcareous mud. Crinoidal limestones are especially prominent in Upper Palæozoic formations.

In certain marine areas enormous accumulations of molluscan shells occur, consisting chiefly of lamellibranchs and gastropods, but also with large numbers of brachiopods and echinoderms, etc., all being more or less broken up. These form the shell sands and shell beaches of exposed shores such as the West of England and of Holland. The floors of shallow seas, as, for example, the North Sea, are largely covered by deposits of shelly sand. Among older rocks consolidated shell beds are frequent. The Pliocene Crags of East Anglia, however, are still in a loose condition. The so-called Purbeck marble is a consolidated mass of shells of a fresh-water gastropod (*Viviparus cariniferus*). It has been largely used as a building stone in the South-west of England.

The calcareous algæ have the power of precipitating

[1] *Geol. Mag.*, 61, 1924, pp. 102-16.

carbonate of lime from bicarbonate solutions. The material is deposited on the " leaves " or thalli of the plants, producing a laminated structure which simulates their general form. This process may take place in lakes with the formation of *fresh water* or *lake marl*, in which the algal aquatic plant, the stonewort (*Chara*), takes a leading part. Fossil Chara limestones are known from the Oligocene (Bembridge, Isle of Wight), and cherts from the Jurassic (Purbeck). The importance of marine calcareous algæ as limestone formers is now widely recognised.[1] The curious layered coralloidal growths commonly known as *Cryptozoon*, which abound in Pre-Cambrian limestones, are now believed to represent algal secretions.[2] Calcareous algæ are very abundant in the Ordovician limestones of Scotland and in the Carboniferous of England. The genera *Girvanella* and *Solenopora* occur in both formations, while *Mitcheldeania* only becomes important in the Carboniferous. The Lithothamnion limestone of the Vienna basin is a Miocene example of an algal limestone. The Lithothamnion sand of the raised beaches of the Clyde has already been mentioned (p. 208). The tiny oval plates (coccoliths) and rods (rhabdoliths) found in oozes and in the Chalk, are the secretions of minute calcareous algæ.

Coral reefs provide the outstanding example of limestones which are built up from the beginning as solid and continuous masses (stereophytic). The coral polypes secrete their skeletal framework in a coherent form which joins up solidly with the rest of the reef, whether that be the original coral growth or a talus of wave-broken fragments over which the reef advances. This type of coral limestone (or coral rock) occurs in structureless, irregular, or lens-shaped masses, distinguished by the absence of stratification from the bio-mechanical limestones which enwrap them. Recent research has shown that other organisms take prominent parts in building coral reefs, especially the calcareous algæ. On an

[1] E. J. Garwood, *Geol. Mag.*, 1913, pp. 440-6 ; 490-8 ; 545-53. Also *Rept. Brit. Assoc. Birmingham*, 1913, Pres. Address, Sect. C., pp. 19. The most complete recent account is that of W. S. Glock, ' Algæ as Limestone Makers and Climate Indicators," *Amer. Journ. Sci.*, 6, 1923, pp. 377-408, which also refers to the precipitation of $CaCO_3$ by bacteria.

[2] Pirsson and Schuchert, *Textbook of Geology*, pt. 2, *Historical Geology*, 2nd ed., 1924, p. 176.

actively-growing reef there is often a raised edge which is composed largely of algæ. Numerous shells and hard parts of other organisms which live on, and in, coral reefs are accidentally involved in the coral growths.

Coral reefs are recognisable in various parts of the stratigraphical sequence. The steep-sided " knolls " of massive limestone in the Carboniferous Limestone of Yorkshire were regarded by Tiddeman as fossil coral reefs. This interpretation is supported by Professor E. J. Garwood and Miss Goodyear for the most typical reef knolls, with the qualification that their inverted basin structure may have been accentuated by subsequent crustal movement.[1]

Much of the powdery calcareous ooze which is found on some tropical sea floors (as, for example, off Florida) is of biochemical origin, and has been precipitated by the nitrifying bacteria which swarm in these seas. The metabolism of these organisms generates ammonium carbonate, which precipitates lime from sea water.[2]

PHOSPHATIC DEPOSITS OF ORGANIC ORIGIN [3]—The ultimate source of the phosphoric acid present in phosphate rocks is the apatite of igneous rocks. This mineral is broken up by mechanical weathering and slowly dissolved by carbonated waters, the solutions being carried by rivers into the sea. Phosphoric acid is present in sea water to the amount of ·015 per cent., or ·18 per cent. of the sea salts. As calcium phosphate it is utilised by certain organisms, especially fish, crustacea, and some brachiopods, in building their shells and skeletons. The remains of these creatures accumulate on the sea floor and form weakly phosphatic deposits, which must be concentrated in various ways in order to form the phosphates of commercial value. Certain phosphatic chalks and limestones may be fossil representatives of these primary phosphatic deposits.

[1] *Quart. Journ. Geol. Soc.*, 80, 1924, p. 258.

[2] G. H. Drew, *Carn. Inst. Washington Publ.* 182, 1914. See also W. S. Glock, *op. cit.*, pp. 379-85. Considerable doubt has been thrown on Drew's bacterial hypothesis by C. B. Lipman, *Carn. Inst. Washington, Publ.* 340, 1925, pp. 181-91.

[3] A recent account of the geology of phosphates is given by Professor J. W. Gregory, *Trans. Geol Soc. Glasgow*, 16, pt. 2, 1917, pp. 115-63. The chemistry of phosphate rocks is dealt with by F. W. Clarke, *Data of Geochemistry*, 5th ed., 1924, pp. 523-34.

Concentration may be effected by the re-solution of the phosphate, and its colloidal deposition around nuclei to form nodular concretions, from which the finer non-phosphatic material may be washed away. Thus are explained the phosphatic nodules scattered abundantly over the ocean floor in many places. The Algerian phosphatic chalks, and beds of phosphatic nodules like those of the Cambridge Greensand, may have been formed in this way. After the sediments have been uplifted, further concentration may be effected by the leaching of the more soluble carbonate materials by surface waters, leaving the less soluble phosphates as a residual deposit. Thus, it is thought, many of the " rock phosphates " of the United States have been formed.

Guano is another phosphate deposit of directly organic origin. It represents a process of concentration which is completed on land. Fish-eating sea birds live in great flocks on small islands for safety, and the guano of commerce is obtained from accumulations of their excrement, which is very rich in nitrogenous and phosphatic material. As guano is a very soluble substance, it is only preserved on islands in regions which are practically rainless. Thus this valuable manure is limited to certain islands off the western coasts of South America, South Africa, and Australia, the main supply coming from islands off the Peruvian coast. Guano is a soft, loose material; but the percolation of a small amount of water is sufficient to render it granular and oolitic through solution and colloidal re-deposition. In less arid climates the nitrogenous matter of the guano is removed, and a less soluble phosphatic residue is left, which is known as leached or phosphatic guano.

The drainage of phosphate-rich solutions from guano deposits into the subjacent rock leads to the metasomatic replacement of the rock by phosphates. Thus phosphatised limestone is formed from coral-rock, as on Christmas and Ocean Islands. On Clipperton Atoll a trachyte has been converted into aluminium and iron phosphates, whilst still retaining its original texture. The Florida rock phosphate may be due to the phosphatisation of limestone islands under caps of guano.[1]

[1] J. W. Gregory, *op. cit.*, p. 137.

FERRUGINOUS DEPOSITS OF ORGANIC ORIGIN [1]—The conditions under which organisms are capable of precipitating
ferruginous materials interlock very closely with those under
which purely chemical action is responsible for the precipitation, and it is difficult to separate the respective effects
(p. 229). Ferric oxide and ferrous sulphide are known to be
precipitated by certain lowly organisms of algal and bacterial character. The organisms may depend entirely on
the iron present in solution, as do the iron bacteria ; or they
may depend for their existence on the supply of carbonaceous
matter, oxygen, etc., the precipitation of iron then being
only an incidental by-product of their biochemical activities. The iron bacteria (or algæ) have the power of extracting
iron from solution, and depositing it as ferric oxide around
their cells. Other organisms may simply act as gathering
surfaces or media for iron oxides precipitated by purely
chemical processes. The accumulation of the ferriferous
casts of the bacteria, along with granules due to chemical
precipitation, produces the material known as *bog iron ore*.
In the case of fossil bog iron ore deposits it is very difficult
to estimate the relative parts played by organisms and
chemical agencies, as the bacterial remains are only recognisable in freshly-deposited material, and are destroyed by
the slightest alteration. Bacteria and other organisms were,
no doubt, indirectly active in the precipitation of the marine
chamositic iron ores, by yielding metabolic products favourable to their deposition.

Where iron salts, sulphates, and sulphuretted hydrogen
(derived from decaying organic matter) are simultaneously
available in solution, as in enclosed bodies of water like the
Black Sea, ferrous sulphide may be precipitated either by
the reducing action of the sulphuretted hydrogen upon the
iron salts, or by the activity of sulphur bacteria which obtain
their oxygen from sulphates, sulphides, or thiosulphates.
The sulphuretted hydrogen itself may be produced by bacterial action from organic matter. The precipitated ferrous
sulphide, probably in colloidal condition, is always more or
less mingled with clayey and organic matter, and goes to

---

[1] A. F. Hallimond, *op. cit. supra*, pp. 14 ; 101-3 ; E. C. Harder,
" Iron-depositing Bacteria and Their Geologic Relations," *Prof. Paper*,
U.S. Geol. Surv. No. 113, 1919, pp. 75-84.

16

form a black mud.   The black and grey shales and slates of the geological column, so often rich in pyrites, are believed to be fossil representatives of these muds.

SILICEOUS DEPOSITS OF ORGANIC ORIGIN—The biogenic siliceous deposits include the present-day radiolarian and diatomaceous oozes and earths, with their fossil representatives, and certain rocks composed largely of the remains of siliceous sponges.   Radiolaria are lowly organisms belonging to the Protozoa.   Their siliceous shells, owing to their relative insolubility, can sink to greater depths than those of calcareous organisms, and are therefore associated with the abyssal *red clay*, the ultimate repository of the non-organic materials which arrive in the ocean depths.   When radiolaria form more than 20 per cent. of this deposit, it is called *radiolarian ooze*.   Sponge spicules and diatoms may also be present.   Radiolarian earths occur in Barbados, associated with red clay and Globigerina beds.   These undoubtedly represent oceanic oozes, and have been protected from erosion by the growth of a coral reef over them.

Radiolarian deposits may, of course, also be formed in shallow waters, and many of the *radiolarian cherts* and *radiolarites* [1] which occur at frequent intervals in the geological column, are believed to have been of shallow-water origin. The organisms thrived wherever there was an accession of soluble silica to the sea water, whether by submarine volcanic action or by other means (p. 225).   Many radiolarian cherts are associated with the submarine pillow lavas.   The organisms may be excellently preserved in silica, or may appear as minute rounded bodies composed of cryptocrystalline silica.   The Lower Carboniferous radiolarian cherts of the Gower peninsula (Glamorganshire) are strongly wedge-bedded, and alternate with shales which yield lamellibranchs and plant remains.   They are interpreted as shallow lagoon deposits. [2]

Diatoms are lowly plant organisms which secrete minute, ornamented, spherical, and discoidal frustules composed of silica.   They thrive in both fresh and salt waters.   Marine deposits in which diatoms are the predominant micro-

[1] Grabau, *Principles of Stratigraphy*, 1913, p. 459.
[2] E. E. L. Dixon and A. Vaughan, *Quart. Journ. Geol. Soc.*, 67, 1911, p. 519.

organisms are limited to the subpolar regions, and the oozes in which they form an important constituent are thus well developed in the Antarctic Ocean, and in the Northern Pacific. A considerable amount of terrigenous material is always found, and diatomaceous oozes grade into the ordinary blue muds of the continental shelves and slopes. Whitish or yellowish chalk-like diatomaceous deposits accumulate on the floors of lakes and in swamps of cold temperate regions, and form the materials known variously as *diatomaceous earth*, *tripoli*, or *kieselguhr*. These are of some commercial importance as abrasives and polishing materials, and as absorbents used in the manufacture of high explosives.

CARBONACEOUS DEPOSITS : PEAT AND COAL [1]—In this section are included all modern and ancient deposits in which the most significant constituent is carbonaceous organic matter. Peat, lignite, coal, anthracite, cannel, and boghead belong here. All these rocks consist very largely of plant debris in various stages of alteration.

The greatest development of *peat* is found in temperate and cold regions ; it is rare in the tropics. The largest peat bogs in the world are situated in Arctic Europe, Asia, and North America. In high latitudes these deposits form the tundras, beneath which the subsoil is permanently frozen. In the British Isles two varieties of peat may be distinguished :

[1] From the voluminous literature of coal it is only possible here to select a few of the recent references, to which one or two are added in the text : E. A. N. Arber, " The Natural History of Coal," *Cambridge Manuals*, 1911, 164 pp.; H. Potonié, *Die Rezenten Kaustobiolithe und ihre Lagerstätten*, three vols. ; *Abhandl. K. Preuss., Geol. Landes.*, 55, 1908-12 ; D. White and R. Thiessen, "The Origin of Coal," with a chapter on the "Origin of Peat," *U.S. Bur. Mines, Bull.* 38, 1913, 304 pp.; H. Potonié, *Die Enstehung der Steinkohle und der Kaustobiolith*, Berlin, 1914, pp. 225 ; E. C. Jeffrey, " The Mode of Origin of Coal," *Journ. Geol.*, 23, 1915, pp. 218-30 ; M. C. Stopes and R. V. Wheeler, " Monograph on the Constitution of Coal," *Dept. Sci. and Industrial Research*, London, 1918, 58 pp. ; M. C. Stopes, " On the Four Visible Ingredients in Banded Bituminous Coal," *Proc. Roy. Soc. Lond.*, B. 90, 1919, pp. 479-87 ; R. Thiessen, " Compilation and Composition of Bituminous Coals," *Journ. Geol.*, 28, 1920, pp. 185-209 ; R. Thiessen, " Structure in Palæozoic Bituminous Coals," *U.S. Bur. Mines, Bull.* 117, 1921, 296 pp.; E. C. Jeffrey, *Coal and Civilisation*, 1925, 178 pp. ; R. Thiessen, "Origin of the Boghead Coals," *Prof. Paper* 132—I, *U.S. Geol. Surv.*, 1925, pp. 121-38 ; R. Potonié, *Einführung in die allgemeine Kohlenpetrographie*, Berlin, 1924, 285 pp.

hill or upland peat, and fen peat or swamp muck. Upland peat is a brownish, fibrous, spongy substance still clearly of vegetable origin, composed of the remains of *Sphagnum* and other mosses, in which the surface layer consists of living plants. Trunks and branches of trees may be incorporated in peat (bog-oak of Ireland), and it may also be mingled with more or less inorganic mineral matter. Peat is due to the gradual accumulation of the dead vegetable fibres as the living upper bed dies and gives place to another layer. In places the thickness of the deposit may reach 50 feet. Its growth is a slow process dependent on climate. In many British localities the growth of peat is now almost or quite stationary. In Scottish peats Lewis has established nine different alternations of bog floras and forest floras, the latter indicating periods of milder climate.

Fen peat is of darker hue and has more of the nature of vegetable muck than hill peat. It consists chiefly of the remains of sedges, rushes, and water-plants, and may contain trees and branches, fresh-water shells, and other organisms. Buried forests are common at definite levels, indicating periods of varying climate or of more effective drainage. The fine, black, lake and swamp muck described by Jeffrey [1] and others, seems to be of the same character as fen peat. It consists of a heterogeneous macerated mixture of the remains of water-plants, amphibious vegetation, tree-trunks and branches, leaves, pollen, and charred wood from forest fires.

Lignite or *brown coal* is more solid than peat, but less compact than ordinary coal. It may be of any tint between brown and black, and retains a distinct fibrous woody structure. It represents an intermediate stage between wood and coal. Great lignite deposits of Kainozoic age occur in Europe, and of Cretaceous and Kainozoic age in North America. The chief example in Great Britain is that of Bovey Tracy, Devonshire, where thick lignite beds are associated with sands, clays, some torrential deposits, and re-deposited china clay, and are probably of detrital origin.

The common *humic* or *bituminous coal*, including house coal, coking coal, steam coal, etc., consists of a stratified,

---

[1] *Coal and Civilisation.* 1925, Chap. III.

compressed, and altered mass of all kinds of vegetable matter in various stages of preservation. It is well jointed, breaking into rectangular lumps, and is often finely laminated. Two textural varieties of coal are commonly recognised : mineral charcoal, mother of coal, or *fusain*, the soft, pulverulent, charcoal-like substance which is found sparingly on certain stratification planes ; and compact coal. The latter, on close examination, shows two kinds of layers : *bright coal* (glance coal), hard, with a jet-black pitchy lustre and conchoidal fracture ; and *dull coal* (matt coal), greyish-black, with a dead lustreless appearance and rough fracture. The bright coal always occurs as lenticular masses, often much elongated, in a matrix of dull coal. With a lens the dull coal is found also to be extensively sublaminated with microscopic streaks of bright coal within a dull structureless matrix. According to Thiessen [1] the bright coal layers are derived from the woody parts of plants (*anthraxylon*), the thicker bands representing trunks and large branches, the thinner ones small branches and twigs. Rectangular chips of wood, such as are found in the litter and peat of present-day forest swamps, form the microscopic streaks of anthraxylon in dull coal, the matrix of which is composed of a large number of other plant constituents. In thin section it appears as a granular mass which, at high magnification, is seen to consist of the waxy cuticles of leaves, spore coats or exines, including both megaspores and microspores, exines of pollen grains, resinous matter, bark, opaque black carbonaceous matter of uncertain origin, and humic matter comprising the cellulosic products of vegetable decomposition. The recognisable fragments thus consist of the more resistant parts of plants. Each has a more or less definite microscopical appearance. These constituents vary in amount along with anthraxylon in different kinds of coal.

Dr. M. Stopes [2] has distinguished four ingredients in bituminous coals, which may be recognised by their macroscopic structures, their microscopic features, and their reactions with chemical agents. *Fusain*, or mineral charcoal, is one of these constituents ; dull coal, or *durain*, is another. The

---

[1] *U.S. Bur. Mines, Bull.* 117, 1920, pp. 22 *et seq.* Also *Journ. Geol.,* 28, 1920, pp. 185-209.

[2] *Proc. Roy. Soc. Lond., B.* 90, 1919, pp. 470-87.

bright or glance coal of other investigators is divided into *clarain* and *vitrain*. Clarain has a smooth surface and glossy lustre which shows fine lamination; thin durain bands are also intercalated. It is the coal material which is richest in recognisable plant matter. Vitrain is a perfectly homogeneous material of brilliant lustre and conchoidal fracture,

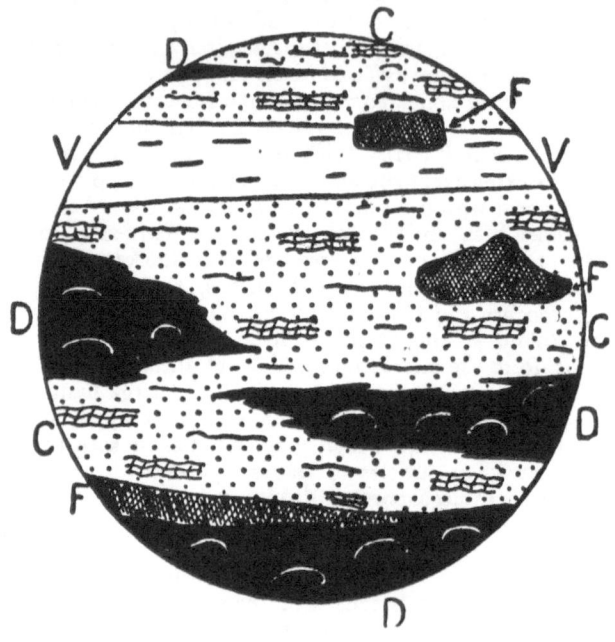

FIG. 64.—MICROSCOPIC STRUCTURES OF COAL.
F, fusain; D, durain; C, clarain; V, vitrain. (For explanation see text, p. 245.) After Dr. M. Stopes.

which presents straight clear-cut margins to the other constituents. Clarain and durain tend to interlock in fine laminæ at their mutual contacts; while fusain forms irregular patches and wedges (Fig. 64). Vitrain seems to represent the hardened colloidal carbonaceous jelly resulting from complete decomposition of plant matter.

Anthracite is a hard variety of coal with a semi-metallic lustre and conspicuous conchoidal fracture. Of all coals it is the richest in carbon and the poorest in volatile materials. Turner and Randall,[1] by means of a new method of etching polished surfaces with a blowpipe flame, have demonstrated that anthracite consists of precisely the same constituents as ordinary coal. A. Stuart,[2] working on the anthracite of South Wales, has been able to distinguish bright black bands of splendent lustre and structureless character, corresponding to vitrain; black bands of fairly bright coal with waxy lustre, and dull black " charcoal " bands, representing fusain.

Cannel coal, with *boghead* or *torbanite*, is a dull, black, compact variety of coal, with a pitch-like appearance and conchoidal fracture. The name cannel refers to its property of easy ignition and burning with a bright, smoky flame like a candle. Cannel coals generally contain a considerable amount of mineral matter or ash, and when the mineral matter so increases as to predominate over the carbonaceous matter, cannel passes into oil shale (p. 215). Cannel coals contain humic, spore, waxy, and resinous plant matter, with the spore and resinous material usually in greater relative abundance than in ordinary coal, mingled with a considerable amount of structureless material. Potonié regarded cannel as due to the solidification of a jelly-like carbonaceous slime (*sapropel*). In thin section boghead coal shows numbers of peculiar irregular, oval, yellow bodies, which have been interpreted as algæ by some writers, and as spores by others. The theory of algal origin suffered from the fact that no known algæ had anything like the requisite chemical composition for the formation of boghead coal, a material which, when distilled, is richly productive of oil. The authors of the theory had to assume that the algæ had an " affinity " for bituminous substances, and had been thoroughly impregnated with, and enriched by, them. The algal theory has, however, recently been rehabilitated by the discovery of richly oil-bearing algæ which live in salt lagoons (South Australia, Turkestan). These form a scum which is blown at times upon the land,

[1] *Journ. Geol.*, 31, 1923, pp. 306-13.
[2] *Geol. Mag.*, 61, 1924, pp. 360-6.

and there consolidates to masses of a rubber-like substance called Australian caoutchouc or coorongite.[1] In thin section this material shows identical appearances with those of torbanite. Coorongite may thus be regarded as the peat stage of boghead coal. Many cannel coals, however, are rich in undoubted spore material.

In discussing the origin of coal, two distinct questions, one geological and the other chemical, arise ; the mode of accumulation of the vegetable matter, and its transformation into coal. As regards the former, the opposition between the growth-in-place and the drift theories of the mode of accumulation, is alluded to in many geological textbooks. The majority of geologists believe that the coal vegetation was fossilised practically on the site of growth. Thiessen states that the origin of peat must be analogous to that of coal, because of the complete congruity of their constituents, and he is convinced that coals had their origin in peat beds.[2] The drift theory, however, is strongly held by other geologists, and there can be no doubt but that some coals have been formed by detrital deposition. The idea of the formation of coal from the accumulations of vegetable matter in forest swamps, like the mangrove swamps of tropical regions, the Everglades of Florida, and the Great Dismal Swamp of Virginia and North Carolina, partially reconciles the two theories, as *in situ* accumulation would occur on relatively-dry areas, and drift accumulation in the numerous open-water spaces.

Chemically, vegetable matter consists of carbon, hydrogen, oxygen, nitrogen, combined in various ways, with certain mineral substances which constitute the ash. Its transformation into coal involves the gradual elimination of the volatile constituents, chiefly as water, but also as marsh-gas and other gases, with a progressive increase in the proportion of carbon. The chemistry of the change is not yet well understood. There is undoubtedly first a biochemical process, whereby the plant material is partly or wholly broken down by the attack of micro-organisms, especially bacteria. This process goes on for varying lengths of time, with varying degrees of intensity ; it may be completely

[1] *U.S. Geol. Surv., Prof. Paper*, 132, I, pp. 121-35.
[2] *U.S. Bur. Mines, Bull.* 117, 1920, p. 13.

arrested if the material is covered with aseptic peat water, or sealed up by sediment.   Hence the original peat may vary in composition from practically undecomposed compressed plant materials to rotted and disintegrated vegetable slime in which little organic structure may be preserved.

A dynamochemical process follows when the vegetable matter is covered with a considerable thickness of sediment, and is thus subjected to pressure and heat.   Spring has shown by experiment that peat, subjected to a pressure of 6000 atmospheres, is converted to a coal-like substance.[1]   Under these circumstances there is a progressive elimination of water and other gases, with a storing up of carbon.   The series peat, lignite, bituminous coal, anthracite, may represent successive stages of this process.   Cannel coals and torbanites are formed by similar processes acting upon spore-rich vegetable slime, or on accumulations of oil-rich algæ.

[1] *Bull. Acad. Roy. Bruxelles*, 49, 1880, p. 367.

PART III

THE METAMORPHIC ROCKS

CHAPTER XV

METAMORPHISM

GENERAL CHARACTER OF METAMORPHISM [1]—Metamorphism may be defined as the response in solid rocks to pronounced changes of temperature, pressure, and chemical environment, which take place, in general, below the shells of weathering and cementation. In the great majority of cases metamorphism means the partial or complete recrystallisation of a rock, and the production of new structures; but the above definition does not exclude the effects of simple crushing and mylonisation with little or no recrystallisation.

Changes in the three factors of temperature, pressure, and chemical environment, upset the physical and chemical equilibrium of a mineral assemblage, and metamorphism results from the effort to establish a new equilibrium. The

[1] For metamorphism in general, see the following works in addition to the textbooks of general petrology: F. Becke, *Über Mineralbestand und Struktur der Kristallinischen Schiefer. Congr. Géol. Internat.*, Wien, 1903, pp. 553-70; C. R. van Hise, "A Treatise on Metamorphism," Mon. 47, *U.S. Geol. Surv.*, 1904, 1286 pp.; V. M. Goldschmidt, "Die Gesetze der Gesteinsmetamorphose," *Videnskapsselsk, Skr.* 1, *M.-N. Kl.*, Kristiania, 1912, No. 22, 16 pp.; C. K. Leith and W. J. Mead, *Metamorphic Geology*, 1915, 337 pp.; R. A. Daly, "Metamorphism and Its Phases," *Bull. Geol. Soc. Amer.*, 28, 1917, pp. 375-418; A. Harker, "Metamorphism in Rock Masses," Pres. Address, Geol. Soc., *Quart. Journ. Geol. Soc.*, 74, pt. 1, 1919, *Proc.*, pp. lxiii-lxxx; L. Milch, "Die Umwandlung der Gesteine." In *Grundzüge der Geologie*, Bd. 1, Teil 1, 1922, pp. 267-327; U. Grubenmann, *Die Kristallinen Schiefer*, 1910; U. Grubenmann and P. Niggli, *Die Gesteinsmetamorphose*, 1, 1924, 539 pp.

processes of weathering and cementation, which take place at nearly normal surface temperatures and pressures, are excluded from metamorphism except by those geologists (of the school of Van Hise) who define metamorphism as equivalent to simple alteration.   At the other extreme, general melting and mixing (anatexis, ultrametamorphism, palingenesis) may occur, and may be included as metamorphism so long as the identity of the altered rock is not completely destroyed by merging into a new magma.

By metamorphism the constituent minerals of a rock are changed over to others which are more stable under the new conditions, and these may arrange themselves with the production of structures which are likewise better suited to the new environment.

AGENTS OF METAMORPHISM—Metamorphism is due to the operation of the three factors of temperature, pressure, and chemically-active fluids. As metamorphism takes place in depth it follows that it depends upon augmented temperatures and pressures, and the increased activity of fluids. The heat may be supplied from the general increase of temperature downward, or from contiguous magmas. The pressure is due ultimately to gravity, and may be resolved at any point into two kinds : hydrostatic or uniform pressure, which leads to change of volume ; and non-uniform, directed pressure, or stress, which leads to change of shape or distortion.   The metamorphic results of these kinds of pressure are very different. While uniform pressure may be applied to liquids or solids, directed pressure can only exist in solid or quasi-solid material.

The environment of chemically-active fluids is a most important factor in metamorphism, as the reactions involved can only take place through partial or complete solution of the minerals. The universal vehicle of alteration is the interstitial volatile or liquid matter which occupies the numberless capillary pores and fissures. Water is undoubtedly the chief substance present in this way ; but it may be reinforced locally by carbon dioxide, and by substances, such as boric and hydrofluoric acids, which emanate from igneous magmas.

The greater part of the volatile substances present in plutonic magmas must ultimately pass into the surrounding rocks ; and their diffusion, although extremely slow, takes them into regions of the crust free from signs of igneous activity. The widespread occurrence of tourmaline in the schists of the Scottish Highlands, even in areas remote from the granite masses which intersect these rocks, points to their thorough permeation by boric acid vapours. Rock moisture (including liquids and gases) then forms a universal medium, extremely dilute and tenuous in general, but attaining a greater concentration and greater chemical activity in the neighbourhood of a plutonic mass, through which mineralogical transformations can take place when and where temperature and pressure conditions are suitable for metamorphism.

The three above-mentioned factors co-operate in various degrees in the production of metamorphism. The chemical factor is, no doubt, at work in practically every case ; but while the combined effects of heat and pressure cannot be dissociated, heat is sometimes the predominant agent of metamorphism, and sometimes pressure.

KINDS OF METAMORPHISM—The different kinds of metamorphism are thus due to various combinations of four agencies : heat, directed pressure, uniform pressure, and chemically-active fluids. The last-named is essential to all kinds of mineral change in metamorphic rocks, with the exception of pure temperature or pressure inversions, such as tridymite to quartz.

1. *Heat Predominant.*—Heat is the dominating factor in the metamorphism which ensues in the proximity of igneous masses. Nevertheless, the factor of pressure comes in through the crowding aside of the country rock, its volume expansion due to heating, and the downwardly-acting weight of the superincumbent mass, although pressure effects are always subordinate to those of heat. In general, the country rock is soaked with gaseous and liquid emanations from the magma, and these greatly facilitate the mineral transformations which take place. The term *thermal metamorphism* may be used for all kinds of change in which heat is the dominating factor.

The term *pyrometamorphism* (Brauns) may be used to denote high temperature changes which take place along the immediate contacts of magma with country rock, and in xenoliths, with or without interchanges of material. The indurating, burning, and fritting effects produced by lavas and small dykes on the rocks with which they come into contact, may be comprised as *caustic metamorphism* (Milch), or better, *optalic metamorphism* (see p. 301).

The general metamorphism which occurs around large igneous masses takes place at comparatively low temperatures (as compared with pyrometamorphism), and is called *contact metamorphism*. In normal contact metamorphism there is little or no change in the bulk composition of the rock. The magmatic emanations in this case merely increase the molecular mobility of the interstitial solutions, and thus facilitate the mineral changes. But in other cases there is a positive increment of magmatic substances to the rock, whereby its composition is altered. This type of alteration is known as *additive* or *pneumatolytic metamorphism ;* and as *injection metamorphism* when there is a more copious addition of magmatic substances which combine with certain excess materials in the metamorphosed rock. Injection metamorphism insensibly passes into that more deeply-seated metamorphic phase, in which the igneous magma itself, or its residual liquid, is bodily injected along the bedding or foliation planes of the affected rock (lit-par-lit injection ; composite- or polymetamorphism). In this, however, pressure may be an important factor.

2. *Directed Pressure Predominant*—A natural pressure is, in general, resolvable into hydrostatic pressure and stress, the latter operating in a definite direction. In co-operation with heat stress is a principal factor in metamorphism. With little or no heat, directed pressure produces crushing and granulation through the enforced movement of rock masses upon one another. This action must be more or less superficial, as in depth heat becomes more and more an operating factor, with the general softening of the rocks, and the increased activity of solutions, leading finally to metamorphism by recrystallisation. The action of directed pressure alone, leads to the mechanical breaking-down of rocks (*cataclasis*), with little new mineral formation except along planes of intense shearing, and with the frequent production of parallel banded structures. Along planes of internal movement frictional heat may be locally developed of sufficient intensity to frit or even melt the adjacent par-

ticles (trapshotten gneiss, flinty crush rock, pseudo-tachylyte, p. 287). The metamorphism which ensues through the dominant action of stress may be called *cataclastic metamorphism*.

3. *Directed Pressure and Heat*—The combination of directed pressure and heat, operating at some depth in the crust, is one of the most powerful in producing metamorphism. It leads to the more or less complete recrystallisation of rocks, combined with the production of new structures. Under these conditions directed pressure lowers the melting-points of minerals locally and temporarily, and thus facilitates diffusion and recrystallisation of material. The main direction of pressure and movement is tangential to the earth's surface. This kind of metamorphism may be called *dynamic*, or better, *dynamo-thermal metamorphism*. The term dynamic metamorphism is defined by Daly [1] as the metamorphism which is induced in rocks in consequence of their deformation, the crustal movements involved being of the orogenic type. It would thus include cataclastic metamorphism as defined above. Dynamo-thermal metamorphism normally takes place in fold-mountain (orogenic) regions, and produces the typical metamorphic rocks, schists and gneisses.

Load metamorphism is a term introduced by Milch [2] to denote the metamorphism primarily due to the vertically-acting stress of superincumbent rocks, aided by the high temperature appropriate to the depth at which it takes place, and by chemical agencies.

Static metamorphism (Daly) is the opposite to dynamic metamorphism, and is defined negatively as that phase of regional metamorphism which is not induced by orogenic deformation.[3] As used, it is nearly synonymous with load metamorphism. Daly divides it into *stato-hydral metamorphism*, taking place at low temperature and in the presence of water, and including such processes as lithifaction and cementation; and *stato-thermal metamorphism*, which is strictly synonymous with load metamorphism, connoting vertical stress and high temperature.

[1] *Op. cit. supra*, p. 397.
[2] *Neues Jahrb. f. Min.*, Beil.-Bd. 9, 1894, p. 121.
[3] *Op. cit. supra*, p. 398.

4. *Uniform Pressure and Heat*—This combination of factors dominates conditions of metamorphism at depths in which directed pressure diminishes and finally ceases to operate, because of the increasing plasticity of the rocks. Mineral transformations are carried to completion, but new structures are not greatly in evidence. Those minerals are formed by which the rock material can best adapt itself to reduced volume ; and they crystallise contemporaneously in an evenly-granular manner, giving rise to the metamorphic types known as granulites. The factors of depth, uniform pressure, and great heat, suggest the term *plutonic metamorphism* as a suitable name for this type of change.

Some geologists have made a fundamental distinction between *local metamorphism* and *regional metamorphism*, the former genetically connected, the latter unconnected or connected only incidentally, with the eruption of magma. If, however, as Harker says, metamorphism is to be envisaged as a single problem, that of the reconstitution of rock masses under varying conditions of temperature and pressure, the above distinction is seen to be unreal and of no practical value.

DEPTH ZONES AND METAMORPHISM—Metamorphism is due to the interaction of three variable factors : heat, directed pressure, and uniform pressure, under the constant activity of interstitial solutions as the media of mineralogical change. Temperature and uniform pressure both increase with depth in the earth's crust ; directed pressure, however, increases with depth only to a certain extent, and then diminishes to zero. On this basis zones of metamorphism correlated with increasing depth have been established. Van Hise [1] thus distinguished a *katamorphic* zone, in general near the surface of the earth, as that region in which destructive alterations take place with the breaking-down of complex minerals into simpler forms ; that is, the region of those disintegrative and decompositional changes which are best exemplified by weathering. Beneath the katamorphic zone lies the zone of *anamorphism*, in which integrating, building-up, constructive alterations take place, involving the formation of complex minerals from those of simpler chemical types. The

[1] *Treatise on Metamorphism*, 1904, pp. 159 *et seq.*

latter are the typical metamorphic changes, and but few geologists follow Van Hise in regarding weathering, cementation, and like processes, as of metamorphic character. The zones of katamorphism and of anamorphism are not sharply separated; on the contrary, they interpenetrate to a large extent, especially in the neighbourhood of igneous masses; and this fact has led Leith and Mead [1] to re-define the terms without reference to depth.

Becke has used the prefixes *kata-* and *ana-* in precisely opposite senses to Van Hise. Grubenmann [2] has also distinguished three zones, an uppermost or *epizone*, an intermediate or *mesozone*, and a lowermost or *katazone*, of metamorphism. Fermor (*Geol. Mag.*, 1927, pp. 334-6) has made the valuable suggestion that the terms *hypozone* and *hypometamorphism* should replace Grubenmann's ambiguous and incorrect terms katazone and katametamorphism.

While the epizone is, in general, nearest to the surface of the earth, the hypozone most remote, and the mesozone of intermediate depth, these zones cannot be said to be sharply marked off from each other, or to have any strict relation to depth. The characters of the zones are determined by the local physico-chemical conditions, which may vary in depth in different places.

FACIES AND GRADES OF METAMORPHISM—The mineral composition of a metamorphic rock is a product of the interaction of two factors, namely, the pressure-temperature conditions under which it was formed, and the chemical composition of the original rock. Grubenmann's zones represent a rough attempt to delimit pressure-temperature conditions, and to classify the rocks accordingly, but there are not nearly enough divisions to express the great diversity of metamorphic types. Eskola [3] has supplemented the zone concept by that of *metamorphic facies*. A metamorphic facies is a group of rocks of varying chemical composition, characterised by a definite set of minerals which have arrived approximately at equilibrium under a given combination of pressure-temperature conditions. Eskola has already distinguished the following facies among the many that still await discrimination and description: hornfels facies, sani-

[1] *Metamorphic Geology*, 1915, p. 19.
[2] *Die Kristallinen Schiefer* I, 1904, p. 55: II, 1907. p. 172.
[3] *Norsk Geol. Tidsskr.*, 6, 1920, pp. 143-94.

Zone.	Temperature.	Uniform Pressure.	Directed Pressure.	Kind of Metamorphism.	Minerals Formed.	Rocks.
EPIZONE .	Low to moderate.	Small.	Often strong; occasionally absent.	Cataclastic to Dynamo-thermal.	Stress minerals (p. 266). Hydroxyl-bearing minerals common.	Phyllites, sericite-, talc-, epidote-, chlorite-, glaucophane-schists. Quartz-schist; schistose-grit, etc.
MESOZONE	Considerable.	Considerable.	Mostly strong.	Dynamo-thermal. Load metamorphism.	Stress minerals predominant.	Mica-schist; garnet-mica-schist, staurolite-schist, hornblende-schist. Mica-, and hornblende-gneiss.
KATAZONE	High.	Very great.	Feeble or absent.	Load or static metamorphism. Plutonic metamorphism.	Anti-stress minerals (p. 263). predominant.	Coarse biotite-, pyroxene-, sillimanite-, and cordierite-gneiss. Granulites, eclogites, etc.

dinite facies, green-schist facies, amphibolite facies, and the eclogite facies.

To the idea of facies, which refers to an assemblage of rock types, Tilley has united that of *grade*, which refers to the stage or degree of metamorphism at which the rocks have arrived.[1] The green-schist facies, for example, is a type of low-grade metamorphism, the eclogite facies of high-grade metamorphism. Rocks belonging to the same facies may be said to be of the same grade of metamorphism. Thus, in the green-schist facies, a chlorite-quartz-muscovite-schist is *isogradic* with a green schist composed of chlorite, epidote, and albite. The latter rock, however, is more susceptible to pressure-temperature changes than the former; and in establishing facies it is preferable to use rocks of this type. The chlorite-quartz-muscovite-schist remains stable over a considerable range of temperature-pressure conditions, and may belong to more than one facies. By selecting one type of rock composition, preferably the most susceptible type, it is possible to map out an area of progressive metamorphism, such as part of the Scottish Highlands, into grades.

Thus Barrow [2] has been able to delimit zones of increasing metamorphic grade in the south-eastern Highlands, which are characterised by the entry of certain index minerals, chlorite, biotite, garnet, staurolite, kyanite, and sillimanite, into rocks of argillaceous composition. We may therefore speak of a chlorite isograd, a biotite isograd, and so on, an *isograd* being defined as a line joining points where the rocks have suffered metamorphism under similar pressure-temperature conditions. An isograd must be the intersection of an inclined isogradic surface with the surface of the earth. Zoning, on this principle, has been carried out in considerable detail by Mr. E. B. Bailey [3] and Dr. C. E. Tilley [4] in the south-western Highlands of Scotland. The conceptions of facies and grade point the way to a possible classification of metamorphic rocks.

[1] *Geol. Mag.*, 61, 1924, pp. 167-71.
[2] *Proc. Geol. Assoc.*, 23, 1912, p. 268.
[3] *Geol. Mag.*, 60, 1923, p. 317.
[4] *Quart. Journ. Geol. Soc.*, 81, pt. 1, 1925, pp. 100-12.

CHAPTER XVI

METAMORPHIC MINERALS, PROCESSES, AND STRUCTURES

INFLUENCE OF ORIGINAL COMPOSITION—The mineral transformations which take place in metamorphism depend, in the first instance, on the composition of the original rock, and then on the kind of metamorphism to which it has been subjected.

Unless there is addition of material from outside, there is no essential change in the chemical composition of a rock during metamorphism, except, perhaps, the partial loss of water and carbon dioxide. A *shale* consisting of quartz, white mica, chlorite, hydrated silicates of alumina, and amorphous iron oxides, will form under contact metamorphism a *hornfels* consisting of quartz, andalusite, cordierite, biotite, and felspar; and under dynamic metamorphism, a *garnetiferous mica schist*, consisting of quartz, muscovite, biotite, and garnet. But the bulk composition of these three types of rock remains substantially identical. The shale mixture is not stable under metamorphic conditions; and the minerals interact with one another, through the medium of the interstitial rock moisture, so as to produce mineral assemblages which are nearer equilibrium under the new conditions of temperature and pressure.

In dealing with metamorphism from the standpoint of initial composition, only four different kinds of original rock materials need be considered. Each group, although composed of rocks of diverse origins, is of substantially the same mineral and chemical composition throughout, and reacts to metamorphism in a distinctive manner:—

1. Argillaceous rocks.
2. Arenaceous rocks, acid igneous rocks and tuffs, acid schists and gneisses.

3. Limestones and other carbonate rocks.

4. Intermediate and basic igneous rocks and their tuffs.

In argillaceous rocks, composed mainly of the finest degradation products of crystalline rocks, and partly of undecomposed rock flour, the constituents are in approximate equilibrium under ordinary surface conditions of low temperature and pressure. Hence, on metamorphism, the successive reactions due to rising temperature or pressure come into play normally and evenly, subject only to the lag effect referred to later. These rocks, therefore, show well-graduated series of changes, and are most suitable for the establishment of successive zones of metamorphism. On the other hand, rocks of the second class, which are composed mainly of quartz and felspars, minerals stable over a wide range of temperature and pressure conditions, only show marked changes at a high grade of metamorphism.

Pure calcium carbonate rocks, again, are stable under metamorphic conditions, and suffer little change except recrystallisation. Dolomite breaks up into calcite and certain magnesian minerals. Impure calcareous and dolomitic sediments, however, as Harker has pointed out,[1] are in a condition of unstable equilibrium, and are extremely ready to fall into new mineral combinations as the conditions of temperature and pressure change. Hence, such reactions, once initiated, go on with a rapidity only limited by the amount of heat available. As they occur chiefly between carbonates and silica, carbon dioxide is released, and is then free to act as a medium in facilitating further changes. Furthermore, as these reactions go on at comparatively low temperatures, they supply the clue to the explanation of the fine grain which is characteristic of many lime-silicate rocks, and also the abrupt transition often seen in the field between altered and unaltered rock, which takes place where the supply of heat has been just insufficient to start the reactions.

The chief minerals of basic igneous rocks, lime-soda plagioclase, pyroxenes, olivine, and iron ores, are fairly susceptible to metamorphic change ; and these rocks are well adapted to exhibit the effects of progressive metamorphism.

Metamorphic reactions go on in the solid state. The rocks

---

[1] Harker, *op. cit.*, p. lxix.

are not melted or dissolved as a whole, and only a small fraction is in solution at any given time. That is to say, the amount of the solid phase in metamorphism is always overwhelmingly in excess of any other phase, liquid or gaseous. This has important effects on the texture of metamorphic rocks; and the diffusion of substance, with the consequent mixing of different materials, must also be extremely limited. This is shown by the frequent preservation of bedding and other sedimentary structures in metamorphic rocks,[1] and by the retention of well-preserved fossils in completely-recrystallised rocks (the Alps, Norway, etc.).

Owing to the conditions of recrystallisation in the solid state, the establishment of equilibrium in metamorphic rocks is often far from complete, and is usually much less rapid than in a freely-mobile magmatic fluid. There is a marked "lag-effect," certain minerals persisting into a region of temperature and pressure wherein they should normally change into a dimorphous form, or react with others to form new minerals. The remains of original minerals that have failed to react to the new conditions, or have become armoured with reaction products, are known as *relicts*. Thus, in the transformation of a gabbro to an eclogite (p. 318), some of the original diallage may escape alteration to omphacite or garnet, and may form a relict mineral. Incomplete reaction is much more common in the metamorphism of a high-grade mineral assemblage, i.e. composed of minerals of complex chemical composition (many igneous rocks, gneisses, etc.), than in the metamorphism of a low-grade assemblage consisting of simple minerals and weathering products, such as clays.

The reverse change is even more difficult. Metamorphism induces a high-grade assemblage of minerals at a high temperature, and on cooling the lag-effect is so great that these minerals in general completely fail to react to the new temperature-pressure conditions. Thus it is, as Harker has shown, that we are able to study high-grade metamorphic rocks at all.[2] Notwithstanding, if sufficient time elapses, regressive changes do occur. If by orogenic movements a

---

[1] "Geology of Ben Wyvis, Carn Chuinneag," etc., *Mem. Geol. Sur. Scotland*, 1912, pp. 75, 78.

[2] *Op. cit.*, p. lxviii.

crystalline schist or gneiss is displaced from a region of plutonic metamorphism to one of meso- or epimetamorphism, it may in time assume the mineral composition and structural characters appropriate to its new environment. Becke has shown that some of the phyllites and fine-grained mica schists of the Alps have been formed from gneisses of the deep-seated zones of the crust. This process is called *regressive metamorphism*.

INFLUENCE OF HEAT AND UNIFORM PRESSURE—Heat and uniform pressure are the dominating factors in thermal and contact metamorphism, and in plutonic metamorphism, although directed pressure is seldom entirely absent. The effects of heat upon minerals and rocks is expressed by the general law of Van t'Hoff, that, at constant volume, rise of temperature produces a displacement of equilibrium towards the absorption of heat; i.e. those minerals will be formed which develop with absorption of heat. Heat also promotes the expulsion of volatile constituents, and increase of volume.

On the other hand, the effect of uniform pressure is to favour the formation of minerals of small specific volume and high density, resulting from the operation of Le Chatelier's law that, if a chemical system be maintained at constant temperature, pressure will displace the equilibrium in the direction of diminution of volume, and in metamorphism will cause reactions which favour the production of high-density minerals. Directed pressure acts in the same way as regards diminution of volume, although it favours the production of quite different minerals.

Those minerals whose formation is favoured by uniform pressure, and which are also well known as products of thermal and contact metamorphism, have been called *anti-stress* minerals by Harker.[1] These minerals, which include anorthite, potash-felspars, augite, hypersthene, olivine, andalusite, sillimanite, cordierite, and spinel, are unstable in the presence of stress.

The relations of heat and pressure in metamorphism can be most simply shown by the pressure-temperature diagram elaborated by Goldschmidt,[2] in which the abscissæ denote pressures, and the ordinates temperatures (Fig. 65). The

---

[1] *Op. cit.*, p. lxxvii.   [2] V. M. Goldschmidt, *op. cit.*, p. 6.

upper limit of metamorphism is taken at the temperature of fusion for most rocks, and is indicated by a broken line start-ing above 1500° C., and rising slightly with high pressure, as pressure raises the melting-points of minerals. Weathering and cementation take place within a small range of tempera-ture and pressure near the point of origin. The temperature and pressure regions characteristic of the epi-, meso-, and

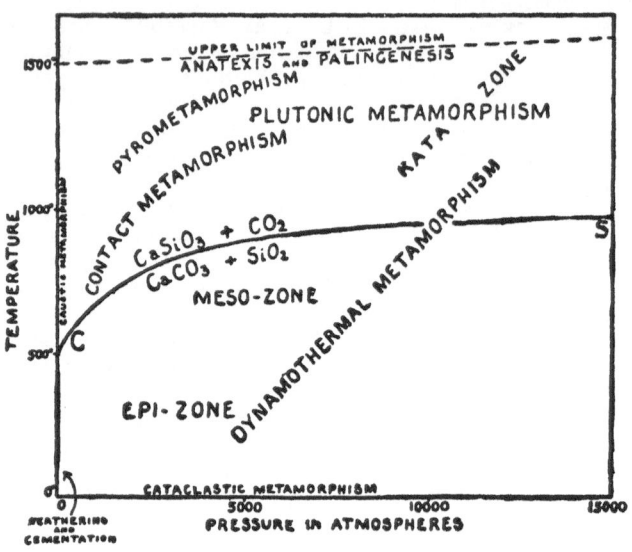

FIG. 65.—PRESSURE-TEMPERATURE DIAGRAM ILLUSTRATING METAMORPHISM.

The diagram shows the fields of the various kinds of metamorphism, and the stability fields of calcite plus silica, and wollastonite. (See text, p. 263.)

kata-zones of metamorphism are roughly indicated, as well as those of cataclastic metamorphism, dynamo-thermal meta-morphism, thermal metamorphism, contact metamorphism, and plutonic metamorphism.

The equilibrium relations of rocks whose components are calcium carbonate and silica are represented by the curve C S. The metamorphism of an important series of sediments, ranging from pure limestones at the one end, through various mixtures of siliceous limestones and calcareous sandstones, to pure sandstone at the other end, can thus be summarised. With rise of temperature calcium carbonate and silica react according to the following equation :—

$$CaCO_3 + SiO_2 = CaSiO_3 + CO_2$$
(calcite)   (silica) (wollastonite) (carbon dioxide)

This reaction is reversible; and the curve shows that heat and pressure act in opposite directions. Increasing temperature causes the reaction to proceed from left to right; increasing pressure, however, reverses the direction. Under pressure-temperature conditions indicated by points below the curve, calcite and silica can exist together; for conditions above the curve they unite to form wollastonite. The reaction depends on the expulsion of carbon dioxide from the calcite, and the curve is drawn through points at which the pressure for any particular temperature is just insufficient to prevent the loss of that constituent. The curve flattens out parallel to the pressure direction, showing that a few degrees rise of temperature is far more effective in promoting a balanced reaction, such as that between calcite and silica, than increase of pressure is in reversing its direction.

Other mineral transformations could be dealt with in the same way were all the necessary data available. While many more data are needed, the stability fields of many minerals and mineral associations are now approximately known.

INFLUENCE OF DIRECTED PRESSURE—Directed pressure or stress is the dominating factor in the epi- and meso-zones of metamorphism, and is mainly concerned in cataclastic metamorphism and dynamo-thermal metamorphism. Stress is an unstable condition, and there is always present a tendency for it to equalise itself and become uniform. The diminution or relief of stress in a rock is accomplished by internal movement and by recrystallisation.

Both the maximum stress and the possible range of stress diminish with increased temperature, and therefore, in general, with depth in the crust. At shallow depths stress acts on hard brittle materials, and manifests itself by crushing, granulation, and shearing in the solid rocks; at greater depths, and with rise of temperature, the rocks become less resistant, and stress is relieved mainly by recrystallisation and flow or fractureless deformation.

The physical chemistry of materials under stress is not so well known as that of materials under uniform pressure, but it is established that directed pressure lowers the melting-points of minerals and increases their solubility. It is, therefore, a potent factor in recrystallisation.

Riecke's principle that solution takes place at the point of greatest pressure in a crystal, with concurrent precipitation at the point of least pressure, has an important bearing on the origin of parallel, schistose, and foliated structures in metamorphic rocks. Crystals under unequal pressure will, in general, grow in a direction perpendicular to that of greatest pressure, with the production of elongated forms orientated parallel to the direction of least pressure. Under directed pressure local melting and solution must take place ; but as stress can only act on a solid phase, the minute quantities of liquid formed are released from stress, and resolidify in a form which is stable under those conditions. The products of the solution will often be new minerals, and as the process goes on the rock finally attains a mineral composition and structure which stress is incapable of modifying further.

The directional element in the pressure also tends to promote mechanical movements in the rocks, partly by the rotation of lath-shaped or prismatic crystals (slaty cleavage), partly by plastic flow or deformation, which takes place chiefly by movement along cleavage and gliding planes in the minerals. In general the minerals will be oriented perpendicular to the direction of greatest pressure. Hence directed pressure is the dominant cause of the parallel structures and textures which are so characteristic of metamorphic rocks in general. Shearing stress favours the production of minerals of the mica group, sericite, muscovite, and chlorite, with albite among the felspars, minerals of the epidote—zoisite group, the amphiboles, along with kyanite, staurolite, chloritoid, and talc, all of which are, therefore, grouped together as *stress minerals.*

TEXTURES AND STRUCTURES OF METAMORPHIC ROCKS [1]—

[1] Holmes, *Petrographic Methods and Calculations*, 1921, pp. 372-83 ; L. Milch, *op. cit. supra*, pp. 310-15 ; Grubenmann and Niggli, *op. cit. supra*, pp. 413-77 ; Becke, *op. cit. supra*.

The distinction between texture and structure remains the same as in the igneous rocks (p. 33) ; but it is harder to distinguish between the two in the case of metamorphic rocks, as different textures may be closely interwoven, and relic textures of the original rocks may be preserved in the midst of textures due to recrystallisation. The textures of metamorphic rocks depend on the shapes of the minerals, and on their modes of growth and mutual arrangement ; their structures depend on the interrelations of various textures within the same rock unit, and are frequently dominated by the directive forces due to unequal pressure.

SHAPES OF METAMORPHIC MINERALS—Many minerals have a natural elongated or flattened habit of growth, and possess one or more good cleavages parallel to the dimensional directions. Some of these minerals, such as the micas, chlorite, talc, and hornblende, are favoured by metamorphic processes. Leith and Mead have attempted to show that there is a marked tendency for rocks rich in these minerals to be developed by metamorphism from several types of sedimentary and igneous rocks ; and that there is in fact an actual change of composition, with a tendency to the elimination of such constituents as are not necessary for the formation of the typical metamorphic minerals.[1] The flaky and rod-like mineral forms are those best adapted to the physical environment produced under metamorphism.

The micas, along with chlorite and talc, are flaky or lamellar in their crystal habits, being well developed in two dimensions, and but poorly in the third. The ratio between the length and thickness of mica flakes (index of elongation) provides a rough index to the intensity of recrystallisation. Thus Trueman has shown that in the biotite of granites the index of elongation averages 1·5, in gneissose granite, 2·5, in contact-metamorphic biotite-schist, 5, and in dynamically-metamorphosed mica schist, 6 to 7. The crystals generally lie in the dimension perpendicular to the direction of greatest pressure ; and as their cleavage-planes are parallel to this surface, an abundance of micaceous minerals causes the development of a schistose structure (p. 272), and planes of easy splitting will occur in the same direction. Micas and

[1] *Metamorphic Geology*, 1915, pp. 200-5

related minerals give rise to a tabular, lamellar, or plane type of schistosity (Fig. 66 A).

Hornblendes, on the other hand, are prismatic, rod-like, or columnar crystals, with a strong development in but one dimension. The index of elongation for hornblende in igneous rocks is about 2·5 ; in hornblende-schist it is about 5. The prisms are generally arranged parallel to the direction of least pressure, like a bundle of pencils. In sections perpendicular to the direction of elongation, a hornblende-schist will show an evenly-granular texture ; in other direc-

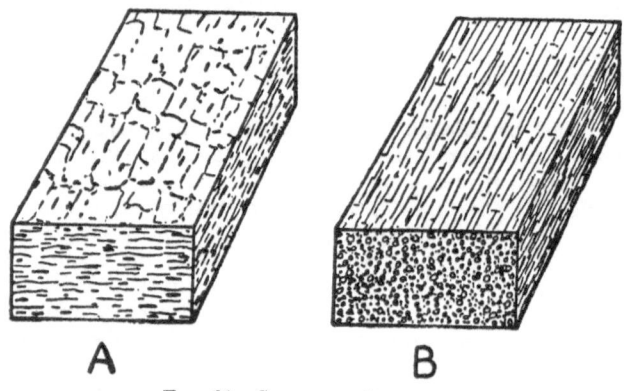

A   B

FIG. 66.—SCHISTOSE STRUCTURES.

Diagrams illustrating tabular and linear schistosity. A, tabular schistosity, due to the predominance of micas or chlorite. B, Linear schistosity, due to predominance of amphiboles. (Text, p. 268.) After Grubenmann-Niggli.

tions the texture will be rod-like or linear. The abundant presence of hornblende or other columnar minerals thus gives rise to a linear type of schistosity (Fig. 66 B).

Most other minerals occurring in metamorphic rocks, such as quartz, felspars, garnets, pyroxenes, and calcite, have, in general, an equidimensional habit. They tend, however, to be somewhat elongated in schists of dynamothermal origin, the leongation being due to plastic flow or to recrystallisation according to Riecke's principle. Hence, in this case a gneissose structure (p. 274) is produced which is due to the parallelism of flattened and lenticular minerals (Fig. 68 c). On the other hand, these minerals remain equidimensional during contact or plutonic metamorphism, and a granulose texture will be produced on metamorphism in rocks mainly composed of them.

GROWTH AND MUTUAL RELATIONS OF MINERALS IN METAMORPHIC ROCKS—As the growth of crystals in metamorphic rocks takes place in a practically solid medium, the textural characters are strongly contrasted with those resulting from crystallisation in the comparative freedom of a magmatic melt. The syllable *blasto-* (from Gr. *blastos*, a sprout) has therefore been used by Becke, either as a suffix or prefix, in the nomenclature of metamorphic textures and structures, to distinguish them from outwardly-similar igneous textures. In many metamorphic rocks the growth of all the minerals takes place almost simultaneously, there is consequently no real order of crystallisation, and the individual minerals tend to enclose each other quite indifferently. Mix-crystals and zoning of crystals are rarely found in metamorphic rocks.

Textures which are due mainly to recrystallisation are described as *crystalloblastic ;* but where original minerals and textures still form an integral part of the rock, the general term *palimpsest* structure [1] is used. In general the proper crystallographic faces are rarely developed in recrystallised metamorphic minerals ; they may be described as *xenoblastic* crystals. A few minerals, however, which possess powerful crystallising force, are able to assert their proper crystalline form, even against the resistance of a solid medium ; these are termed *idioblastic*. Becke has found it possible to give a crystalloblastic series, in which the metamorphic minerals are arranged in order of decreasing tendency to form well-bounded crystals. Thus magnetite, sphene, garnet, andalusite, staurolite, and kyanite, frequently form good idioblastic crystals ; epidote, zoisite, amphiboles, and pyroxenes come next in order ; then come the micas, chlorite, and carbonates ; and finally quartz and felspars, which rarely, if ever, form well-shaped crystals.

Leith and Mead [2] believe that xenoblastic crystals, especi-

[1] Palimpsest, a partially erased manuscript used for later writing.
[2] *Metamorphic Geology*, 1915, p. 187.

ally those of flaky or rod-like forms, are produced during phases of movement in the formation of metamorphic rocks as required by the conditions of unequal pressure ; and that idioblastic crystals are formed in a later static phase, as they often grow quite independently of the schistosity. Thus, in their view there is an order of crystallisation in metamorphic rocks, namely, the xenoblastic crystals first, and then the idioblastic.

For textures produced by recrystallisation (crystallo-blastic textures) the syllable -*blast* is used as a suffix. Thus, when idioblasts form large crystals embedded in a fine-grained groundmass, like the phenocrysts of a porphyritic igneous rock, the term *porphyroblastic* is used to describe the texture. *Granoblastic* indicates recrystallisation tex-tures in which the principal constituents are granular or equidimensional. For palimpsest structures [1] the syllable *blasto-* is used as a prefix. When remnants of original por-phyritic or ophitic textures are recognisable, the terms *blastoporphyritic* and *blastophitic* respectively are applied. When the structure is compounded of recognisable fragments of various types of sedimentary rocks, the terms *blastopsephitic*, *blastopsammitic*, and *blastopelitic* [2] are used for metamor-phosed conglomerates and breccias, arenaceous rocks, and argillaceous rocks respectively.

Structures of Metamorphic Rocks—Holmes [3] has sug-gested a convenient grouping of metamorphic structures into *cataclastic*, *maculose*, *schistose*, *granulose*, and *gneissose*. Cata-clastic structures, as the name indicates, are those of the broken and fragmented rocks developed by shearing stress upon hard, brittle materials in the upper zones of the earth's crust, under conditions involving but little new mineral for-

---

[1] *Structure* is here used as two or more textures are often necessarily in juxtaposition.

[2] The Greek terms *psephitic*, *psammitic*, and *pelitic* are now used as descriptive adjectives for metamorphic rocks derived respectively from gravel rocks, sand rocks, and clay rocks. The corresponding Latin terms rudaceous, arenaceous, and argillaceous, are more familiar, and connote more directly than the Greek terms the characteristic features of the sedimentary rocks themselves. When the sedimentary rocks have been hardened and altered beyond the limits implied by the Latin terms, the less familiar Greek words are more applicable (G. W. Tyrrell, *Geol. Mag.*, 58, 1921, p. 501).

[3] *Petrographic Methods and Calculations*, 1921, p. 372 ; 377-83.

mation (Figs. 67A, 69). Soft rocks like shales or tuffs develop cleavage; harder rocks are shattered and finally crushed to powder, with the formation first of crush-breccias, and at

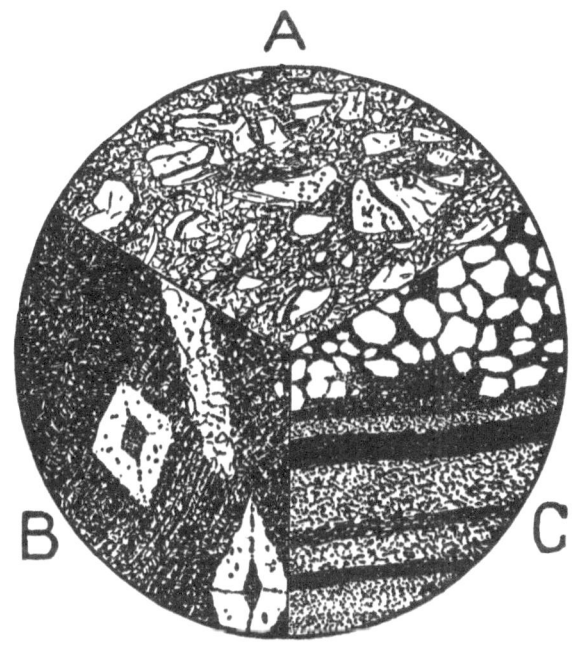

FIG. 67.—STRUCTURES OF METAMORPHIC ROCKS.

A. Cataclastic structure. Cataclasite, Abisko, Swedish Lappland. Due to the crushing of a quartzite. The only new minerals are a few wisps of chlorite.

B. Maculose structure. Chiastolite-slate, Gefrees, Bavaria. Shows porphyroblasts of chiastolite in a fine-grained groundmass in which a strain-slip cleavage has been developed.

C. A palimpsest structure. Garnetiferous biotite-hornfels, from the aureole of the Carn Chuinneag granite-gneiss, Ross-shire. Shows alternations of psammitic and pelitic sediments preserved, although the rock is thoroughly hornfelsed with the production of muscovite and biotite. This rock represents the original materials of the Moine Gneiss. After microphotograph in, *The Geology of Ben Wyvis, Carn Chuinneag*, etc., 1912.

later stages of mylonite (p. 286). The more resistant minerals (e.g. porphyritic felspars), or rock fragments (as in a conglomerate), may be less crushed, and may stand out in a pseudo-porphyritic manner from the finer material produced by the crushing of the softer constituents. This structure is called *porphyroclastic*. Many rocks are drawn out into parallel lenticles, streaks, and bands, of differential crushing and sometimes of different composition, and the structure then simulates that of true schists and gneisses due to recrystallisation. A new term is needed to indicate the pseudo-schistose or pseudo-gneissose structures due to cataclasis.[1] The minerals of cataclastic rocks often show marked strain effects, such as undulose extinction in quartz, and secondary twinning in felspars and calcite.

Maculose structure is that in which porphyroblasts of strong minerals such as andalusite, cordierite, chloritoid, ottrelite, biotite, etc., are well developed, or in which spotting appears as the result of incipient crystallisation of these minerals, and of the segregation of carbonaceous matter. Maculose structure is typically developed in argillaceous rocks under contact or thermal metamorphism. The porphyroblasts form by metamorphic reconstitution from the original mixture of decomposition products and rock flour, which is retained under a low-grade metamorphism as the groundmass in which the porphyroblasts are embedded (Fig. 67 B). The German names, for which there are no good English equivalents, are thoroughly descriptive of these typically spotted rocks; *fleckschiefer*, with minute spots or flecks; *fruchtschiefer*, with spots suggestive of grains of wheat; *garbenschiefer*, with spots resembling carraway seeds; and *knotenschiefer*, with larger spots consisting of individualised minerals which stand out prominently as clots or knots. By continued recrystallisation under more intensive metamorphism these spotted rocks pass over into fine-grained granoblastic types (hornfels), which often show an incipient banding or foliation. From this stage, into which the spotted aspect often persists, there exist transitions to true granulose, schistose, and gneissose structures.

Schistose structure is due to the predominance in a meta-

---

[1] *Cf.* Grubenmann-Niggli, *Gesteinsmetamorphose*, I, 1924, pp. 451, 454.

morphic rock of flaky, lamellar, tabular, rod-like, and highly-cleavable minerals, such as mica, chlorite, talc, and amphiboles, which, under the dominant influence of directed

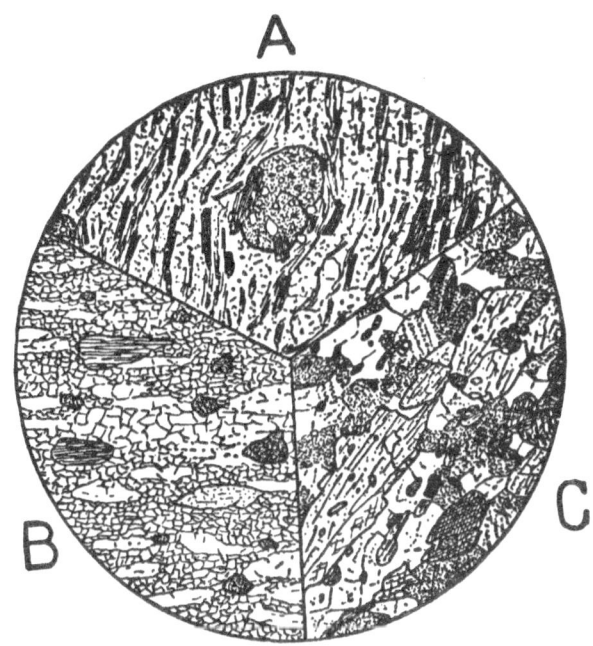

FIG. 68.—STRUCTURES OF METAMORPHIC ROCKS.

A. Schistose structure. Garnetiferous mica-schist, Beinn Achallader, Perthshire. Shows garnet, muscovite, and biotite, in a quartzose groundmass. The structure is determined by the predominance of micaceous minerals. × 25.

B. Granulose structure. Kyanite-garnet-granulite, Rohrsdorf, Saxony. Shows kyanite (cleaved) and garnet (high R.I.) in groundmass consisting mainly of quartz. Structure determined by the predominance of equidimensional minerals. Parallelism due to the flattening of larger quartz grains, and the alternation of streaks of differently-sized grains. × 25.

C. Gneissose structure. Hornblende-gneiss (Lewisian), Loch Glencoul, Sutherlandshire. Shows alternation of folia consisting mainly of quartz and felspars, and of hornblende and epidote. × 15.

18

pressure in dynamo-thermal metamorphism, form layers, felts, and folia arranged in more or less parallel bands. This arrangement of inequidimensional minerals is called *foliation*. The folia may be plane, undulating, or lenticular where they wrap round porphyroblasts ; they may appear megascopically as a continuous felt of flakes or rods, in which case the foliation is said to be closed ; or the dominating minerals may be megascopically discontinuous, and form an open or interrupted foliation.[1] Foliated rocks have the property of schistosity, whereby they can be split along planes parallel to the plane of extension of the constituent minerals, to which the mineral cleavages are often also parallel. In highly micaceous or chloritic rocks the schistosity planes are coated with spangles of mica and chlorite with little or no orientation ; but fracture planes perpendicular to the schistosity show well-developed foliation of a plane or tabular type (Figs. 66 A, 68 A). In hornblendic rocks, however, fracture planes perpendicular to the direction of extension of the prisms will exhibit a granulose texture, while those parallel to it will, in general, show a rod-like or linear type of foliation (Fig. 66 B).

Granulose structure is due to the predominance of equidimensional minerals, such as quartz, felspar, pyroxene, calcite, etc., in a metamorphic rock. The cleavable lamellar or rod-like minerals are either absent or present only in subordinate amount ; and granulose rocks, therefore, do not possess the property of schistosity. The typical texture is coarsely granoblastic ; but parallel banded, streaky, or lenticular structures, may occur from the alternation of patches differing in mineral composition or in grain size (Fig. 68 B).

Gneissose structure is a composite structure due to the alternation of schistose and granulose bands and lenticles, which are dissimilar both in mineral composition and in texture. The foliation is interrupted ; and while a gneiss may split along a plane of schistosity, it does so much less readily than a schist, and exposes a much rougher fracture

---

[1] Closed and open foliation can only be distinguished megascopically. Closed foliation in a hand specimen may or may not be continuous in a microscopic section.

surface. The micas and hornblendes may occur segregated into more or less continuous schist bands or lenticles (Fig. 68 c); or the disconnected flakes and rods may be distributed with parallel orientation through a granulose matrix. In the latter case a banded gneiss or granulite of Moine type (p. 315) is produced. There can obviously be all transitions between schistose, granulose, and gneissose structures.

CLASSIFICATION AND NOMENCLATURE OF METAMORPHIC ROCKS—The classification of metamorphic rocks is attended with serious difficulties. The same initial material may give rise to quite different products under different kinds of metamorphism; and again, practically identical metamorphic rocks may be produced from quite different original materials. Thus, hornblende-schist and amphibolite may be formed by the metamorphism of basic igneous rocks and tuffs, and of certain sediments of mixed composition; and also by the contact action of granite upon calcareous sediments with addition and interchange of material.

The factors which may be used in classification are the initial composition of the metamorphic rocks so far as this can be ascertained, the different kinds of metamorphism, and the grades of intensity of metamorphism. Grubenmann[1] erected his well-known classification of crystalline schists on two of these variables, initial composition and metamorphic grade, the latter being roughly expressed by the triple division into epi-, meso- and kata-zones. The facies classification of Eskola (p. 257) is a more elaborate attempt in the same direction. It seems, however, that in these classifications only those metamorphic rocks in which recrystallisation has been paramount have been treated. Important classes of metamorphic rocks, as, for example, those in which cataclasis is predominant, and those in which addition and interchange of material is a prominent factor, appear to have been left out of consideration. Holmes has proposed a tentative classification in which these defects have been largely remedied,[2] and the classification proposed below represents an expansion of his method.

In the first place metamorphic rocks may be classified

[1] *Die Kristallinen Schiefer*, 1910.
[2] *Nomenclature of Petrology*, 1920, pp. 280-1.

according to whether recrystallisation has or has not been a predominant factor, and whether there has been a significant addition of substance from without. Three main groups may thus be distinguished :—

1. Rocks due to *mechanical* processes. Chiefly mechanical effects (crushing, cleavage, etc.), with a minimum amount of recrystallisation.

2. Rocks due to *recrystallisation* processes. Hornfels, schists, granulites, and gneisses. No serious additions of material.

3. Rocks due to combined *recrystallisation and additive processes*. Impregnated and composite rocks with addition of substances from igneous exudations, and the bodily injection of liquid magmatic materials.

A further division may now be made on the basis of original composition ; and for this the four groups enumerated on page 260 will, in general, be sufficient; although minor compositional groups, such as lateritic, ferruginous, and carbonaceous, may be needed.

Further classification in the second group, and to a less extent in the third, may be made according to structure, which may be construed as roughly corresponding to varying kinds of metamorphism, maculose structure being induced by contact metamorphism, granulose by thermal and plutonic metamorphism, schistose and gneissose, by dynamothermal metamorphism. Finally, the groups thus formed may be subdivided according to grade of metamorphism by the application of the principle of the entry of certain index minerals.

It must not be thought that classification on these lines is at all sharp or definite. All kinds of transition must occur from one group to another, and it may often be difficult to place particular rocks owing to inability to decide how much recrystallisation or additive reaction has taken place, or to determine the nature of the original material.

The nomenclature of the metamorphic rocks has been built up on no definite plan. The terms *schist, gneiss*, and *granulite* are well established for rocks with the corresponding structures ; and *hornfels* is now largely used for the typical products of contact and thermal metamorphism, with maculose structure. To these terms are affixed qualifiers indicating a prominent compositional, mineral, or textural character, as hornblende-schist, cordierite-gneiss, andalusite-hornfels, conglomerate-gneiss, pyroxene-granulite, augen-gneiss, etc. The use of the terms psephitic, psammitic, and pelitic has already been explained (p. 270). The prefixes *para-* and *ortho-* were utilised by Rosenbusch for gneisses derived from sediments and igneous rocks respectively; hence the terms *paragneiss* and *orthogneiss*.[1] These prefixes may be extended with the same signification to schists and granulites. The prefix *meta-* has been extensively used for rocks with palimpsest structures, thus meta-gabbro, meta-basalt, etc. Some metamorphic rocks have the suffix *-ite*, as phyllite, eclogite, mylonite, etc.; while still other names, such as slate, adinole, porphyroid, flinty-crush rock, etc., do not fall within any definite category of nomenclature.

[1] Professor J. W. Gregory has used the term *metapyrigen* gneiss for those rocks of igneous origin on which foliation has been impressed by metamorphism, " The Waldensian Gneisses in the Cottian Sequence," *Q.J.G.S.*, 50, 1894, p. 266.

GROUP I—CATACLASTIC ROCKS

Composition.	Cleaved.	Brecciated.	Phacoidal.	Mylonitic.	Vitrified, or Banded.
Argillaceous .	Slate.				Buchite.
Quartzo-felspathic .	Acid tuff-slate.	Kakirite. Crush-breccia. Crush-conglomerate.	Conglomerate-gneiss. Flaser-granite. Augen-gneiss (part). Porphyroid.	Cataclasite. Mylonite. Mylonite-schist.	Flinty-crush-rock. Ultramylonite. Hartschiefer.
Basic Igneous Rocks	Basic tuff-slate.		Flaser-gabbro.	—	
Calcareous and Dolomitic .	Calcareous slate.		Calc-schist (in part).		—

GROUP II—CRYSTALLOBLASTIC ROCKS

Composition.	Maculose.	Schistose.	Gneissose.	Granulose.
Argillaceous	Spotted slates. Chiastolite-slate. Andalusite-hornfels. Cordierite-hornfels.	Phyllite. Mica-schist. Garnet-mica-schist. Kyanite-schist. Staurolite-schist.	Pelitic gneiss. Para-gneiss. Sillimanite-gneiss. Cordierite-gneiss. Garnetiferous-gneiss.	Kyanite-granulite. Leptite (in part). Kinzigite.
Quartzo-felspathic	Quartzite (in part).	Quartz-schist. Sericite-schist. Schistose grit.	Psephitic gneiss. Psammitic gneiss. Ortho-gneiss. Granite-gneiss. Augen-gneiss (in part).	Quartzite (in part). Granulite. Leptite. Halleflinta.
Basic Igneous Rocks	Pyroxene-hornfels, etc.	Talc-schist. Chlorite-schist. Hornblende-schist. Epidote-schist.	Hornblende-gneiss. Amphibolite (in part). Epidote-biotite-gneiss. Garnet-biotite-gneiss.	Pyroxene-granulite. Garnet-amphibolite (in part). Eclogite.
Calcareous and Dolomitic	Marble (in part). Calc-flintas. Calc-silicate-hornfels.	Calc-schist (in part).	Cipollino. Ophicalcite. Crystalline limestone.	Crystalline limestone, and marble (in part). Dolomite marble. Predazzite and Pencatite. Calc-silicate rocks.

GROUP III—ROCKS DUE TO RECRYSTALLISATION AND ADDITIVE PROCESSES

Composition.	Maculose.	Schistose.	Gneissose.	Granulose.
Argillaceous	Tourmaline-hornfels. Cornubianite. Adinole. Soda-hornfels.	Albite-schist (in part).	Injection-gneiss. Veined-gneiss. Lit-par-lit gneiss. Composite gneiss. Migmatite. Albite-gneiss. (in part).	Migmatised granulites and leptites.
Quartzo-felspathic	Greisen. Quartz-tourmaline and Quartz-topaz rocks.	Tourmaline-schist.		
Basic Igneous Rocks	Skarn.	—	Scapolite-gneiss. Cordierite-anthophyllite rock. Hornblende-gneiss (in part). Skarn-gneiss.	Basic granulite (in part).
Calcareous and Dolomitic	Calc-silicate-hornfels (in part). Garnet rocks.	—	—	Calc-microcline-pyroxene rocks (e.g. Glen Tilt.)

CHAPTER XVII

CATACLASTIC METAMORPHISM AND ITS PRODUCTS

CATACLASTIC METAMORPHISM IN GENERAL [1]—Cataclastic metamorphism results from the crushing and granulation of minerals and rocks (cataclasis), through the application of stress under small load and at low temperatures, with but little new mineral formation, except along planes of con-siderable movement, and at places where heat has been locally generated. At greater depths in the crust, or near the loci of igneous intrusions, where heat becomes a co-operating factor, cataclastic metamorphism passes gradually into dynamo-thermal metamorphism.

Cataclasis may act on fine-grained rock bodies as a whole, producing *crush-breccias ;* or upon individual minerals (in coarse-grained rocks), forming *micro-breccias, flaser-rocks,* and *mylonites.* With simple crushing and granulation a struc-tureless aggregate results ; but when strong lateral move-ment occurs, as along thrust-planes, the broken mineral and rock fragments are rolled out and milled with the production of parallel, lenticular, and banded structures.

Rocks and minerals vary greatly in their resistance to pressure, and their susceptibility to fracture and crushing. At great depths in the crust, where the rocks are confined by enormous hydrostatic pressure and are highly heated, the deformation of all rocks and minerals takes place by plastic flow and recrystallisation. At shallower depths only the softer, more soluble, less brittle rocks and minerals react in

[1] Grubenmann-Niggli, *Die Gesteinsmetamorphose*, I, 1924, pp. 218-30 ; P. Quensel, " Zur Kenntniss der Mylonitbildung," *Bull. Geol. Inst. Upsala*, 15, 1916, pp. 91-116: R. Staub, " Petr. Unters. im westl. Berninagebirge," *Viertelj. Zürich Naturf. Ges.*, 60, 1915, pp. 162 ; P. Termier, *Compt. Rend.*, Paris, 1911, tom. 152, p. 1550.

this way to pressure; the harder and more brittle materials are deformed by fracture and crushing. Hence cataclasis is much more prominent in the hard, brittle, resistant rocks,

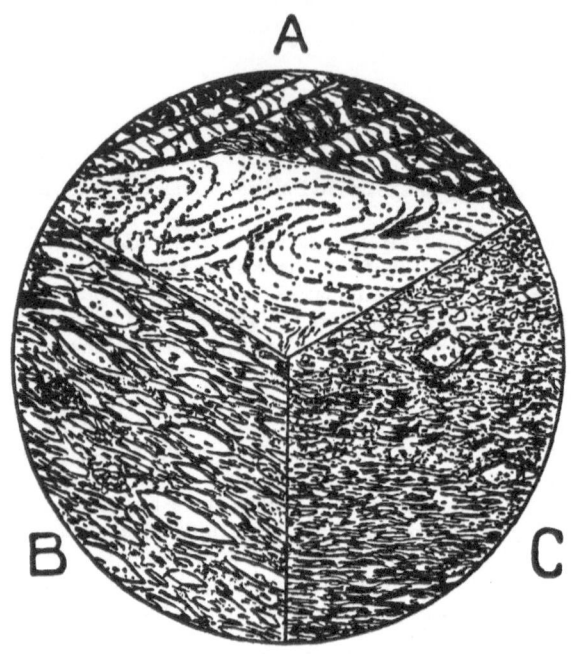

FIG. 69.—CATACLASTIC STRUCTURES.

A. Alternation of fine-grained quartzite and slate, Royal Bay, South Georgia. Shows the different effects of directed pressure on the two rock types. The quartzite has been thrown into sharp folds, while the slate has developed a strain-slip cleavage roughly parallel to the axes of the folds.

B. Flaser structure. Sheared grit, New Fortune Bay, South Georgia. Shows a coarse grit broken down into a flaser structure. Lenticular grains of quartz, and a few of felspar, are embedded in a sheared quartzo-felspathic groundmass, in which a number of shreds of mica have been developed.

C. Mylonite, Abisko, Swedish Lappland. The original syenite (?) has been crushed down and rolled out into a streaky powder. There has been a considerable development of mica in the more felspathic bands.

such as granites and arenaceous sediments, than in the softer and more chemically-susceptible rocks such as the argillaceous and calcareous sediments, and basic igneous rocks. Where rocks of different susceptibility, such as, for example, thin-bedded alternations of quartzite and slate, come under cataclastic conditions, the harder beds yield by folding, fracture, and crushing, while at the same time, the weaker slates are deformed by cleavage, flow, and recrystallisation (Fig. 69 A).

SLATES AND SLATY CLEAVAGE—The chief effect of cataclastic metamorphism on argillaceous rock is the production of *slate*, which splits or cleaves readily along smooth, flat, closely-spaced surfaces of weak cohesion, usually developed at an angle to the original bedding. Slates are mainly composed of finely-divided micaceous minerals, including chlorite, with subordinate quartz and felspars. All the minerals are flattened and elongated in the plane of cleavage.

Slaty cleavage results from the flattening and rotation of mineral fragments, so that the directions of their greatest, mean, and least diameters are brought into parallel position. Under the action of directed pressure the rocks first yield by folding; then, as compression increases, the mineral fragments are gradually rotated into positions perpendicular to the direction of pressure, producing *flow cleavage* (Van Hise and Leith). On the other hand, yielding may sometimes take place along sets of closely-spaced, parallel shear planes which are inclined to the direction of pressure. This gives rise to *fracture cleavage* or *strain-slip cleavage* (Fig. 67 B, 69 A). The cleavage directions must obviously be more or less inclined to the original bedding-planes, which are often obliterated, but can sometimes be recognised as colour bands, or as bands of slightly-differing textures, upon the cleavage surfaces.

As argillaceous materials are especially susceptible to chemical change during metamorphism, a considerable amount of new mineral formation takes place in the transformation of a clay or shale into slate. Hutchings [1] has shown that the recrystallisation of argillaceous rocks begins at a very early stage, as secondary muscovite can be found in soft shale. He also pointed out that the primitive mica

[1] W. M. Hutchings, *Geol. Mag.*, 1890, pp. 264, 316; 1891, p. 164; 1892, pp. 154, 218; 1896, p. 309.

is impure, but sheds its impurities as reconstitution proceeds, while at the same time a chloritic mineral develops. Dr. Brammall has investigated the chemistry of these reconstitution processes in clays, shales, slates, and phyllites,[1] and concludes that they tend to the establishment of a metastable ternary system of white mica, chlorite, and quartz. The change is essentially a process of molecular diffusion, producing a segregation of the monovalent elements combined with alumina, silica, and water, from divalent elements combined with alumina, ferric oxide, silica, and water. The primitive mica thus sheds its impurities of magnesia, lime, and iron oxides, which are taken up by the developing chloritic matter; while the latter loses alkalies and some alumina which supply increment to the developing mica. Quartz is a by-product of these changes, and is also formed by the crystallisation of the colloidal silica of the original clay.

Slaty cleavage is naturally best developed in rocks rich in micaceous minerals. It occurs also in other types of rock, but much less perfectly. In thin-bedded alternations of argillaceous and arenaceous rocks directed pressure may cause cleavage (usually strain-slip cleavage) in the shaly layers, but folding or granulation in the sandstone intercalations (Fig. 69 A).

CRUSH-BRECCIA AND CATACLASITE—If no serious lateral thrusting or shearing takes place during cataclastic metamorphism, the rocks are merely shattered and pulverised, with the formation of a structureless aggregate of fragmental material of various sizes (crush-breccia, kakirite, cataclasite). If, on the other hand, the crushing is accompanied by considerable dislocation and differential mass movement of the material, well-marked lenticular and parallel structures are produced (flaser-rocks, mylonite). The shattering may affect the rock body as a whole, especially fine-grained rocks; or it may affect the individual minerals of coarse-grained rocks, producing brecciation in the one case, and microbrecciation or cataclasis (Daubrée) in the other.

Crush-breccia and *crush-conglomerate* are formed by the mechanical fragmentation, chiefly of hard and brittle rocks. The rock is seamed by fissures in all directions, dividing it

[1] *Min. Mag.*, 19, 1921, pp. 211-24.

up into angular fragments which are separated from each other by pulverised material. The crush-conglomerates of the Isle of Man [1] are due to the shattering of gritty bands intercalated in slate. The slate thus forms the matrix for separated fragments of grit more or less rounded by shearing. In Prince Charles Foreland (Spitsbergen) a thin-bedded sedimentary series consisting of alternations of argillaceous, arenaceous, and calcareous rock-types has been subjected to severe stress, with the result that the more resistant layers of slate and quartzite have been broken into small pieces, whilst the weak calcareous material has been forced to flow between the fragments, and form the matrix of the resulting breccia.[2] The *kakirite* of Swedish Lappland is of the same nature as crush-breccia. The fracture planes are full of rock powder in which there has been a little recrystallisation.[3] When limestones are broken in this way the fissures are often filled with veins of calcite, producing a veined crush-breccia.

The first effect of directed pressure on the larger mineral fragments of a rock is to produce optical anomalies in some ; and bending, or gliding along cleavage planes, in others. Undulose extinction appears in quartz ; a patchy secondary twinning is induced in felspars and calcite ; isotropic minerals such as garnet become irregularly birefringent. The next stage, as the stress exceeds the elastic limits of the minerals, is flattening, elongation, and peripheral granulation. Quartz is more easily broken down than felspars, and the larger crystals become enveloped in a mantle of finely-crushed material (*mortar* structure) ; while felspars are merely flattened and fractured. Hence felspars more often occur as porphyroclasts than quartz. Porphyroclasts which have been crushed into lenticular eye-like forms produce the so-called *augen* structure. The final stage of cataclasis is the complete pulverisation of all minerals ; and when the product is a structureless rock powder, in which a few porphyroclasts may have survived, the term *cataclasite* is used in preference to mylonite, as the latter connotes a rolling out or milling of the material with resulting parallel structure.

[1] G. W. Lamplugh, *Quart. Journ. Geol. Soc.*, 51, 1895, p. 563.
[2] G. W. Tyrrell, " Geology of Prince Charles Foreland," *Trans. Roy. Soc. Edin.*, 53, pt. 2, 1924, pp. 463-4.
[3] P. Quensel, *op. cit. supra*, p. 100.

FLASER ROCKS AND MYLONITE—Rocks in which considerable differential movement has taken place during crushing
exhibit parallel structures in great perfection, and the phenomena of porphyroclasts and augen-structure are even better
developed than in the rocks above-described. Coarse
igneous rocks, such as granite and gabbro, are sheared into
lenticular masses, which are enveloped by streaks of finely-
crushed rock in which a considerable amount of recrystallisation has often taken place. The phacoidal fragments do
not lose all traces of their original characters, and the rocks
as a whole have not been transformed into schists and gneisses.
Similar effects may be produced in acid volcanic or hypabyssal rocks, such as quartz-porphyry or felsite, in which
the quartz and felspar phenocrysts may escape granulation,
and persist as porphyroclasts in a very fine-textured, sheared,
quartz-sericite matrix. These rocks are called *porphyroid*.
The shearing of sandstones and grits has produced some of the
rocks known as *schistose grit* (Fig. 69 B); but many of the
rocks so-called are completely recrystallised. In some sheared
limestones uncrushed lenticles of the original rock have been
preserved in a matrix which, owing to the susceptible nature
of the material, has been largely recrystallised. Rocks in
which lenticles of relatively-unaltered material are preserved
in a finely-crushed and partially-recrystallised matrix, are
called *flaser rocks* (Ger. *flaser* = streaks, lenticles); hence the
terms *flaser-granite, flaser-gabbro*, etc. These rocks form the
" mylonite-gneiss " of Quensel. The new minerals produced
under these circumstances naturally belong to the stress
category. Their mode of formation is more fully treated
elsewhere (p. 308 *et seq.*).

Lapworth's term *mylonite* [1] is applied to rocks which have
been completely pulverised and rolled out (milled) by extreme differential movement during cataclastic metamorphism (Fig. 69 c.) Quensel has used the term for the structureless rock powder here called cataclasite, and has employed
mylonite-schist for pulverised rocks with parallel structures.
The amount of recrystallisation varies with the kind of rock
acted upon. Quartzo-felspathic rocks crush down readily
without much new mineral formation; but basic igneous

[1] *Nature*, 1885, p. 559.

rocks, such as those frequently found on the soles of the thrust-blocks of the north-west Highlands, are sheared into chlorite-schists, in which recrystallisation prevails over cataclasis.

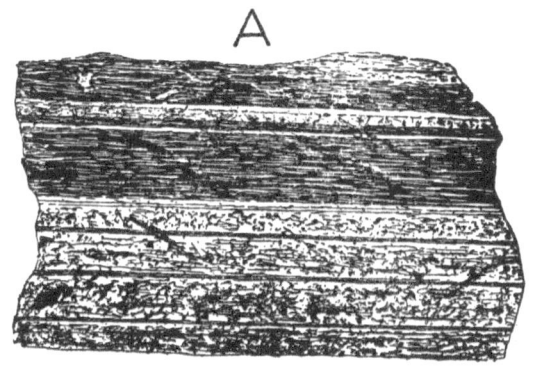

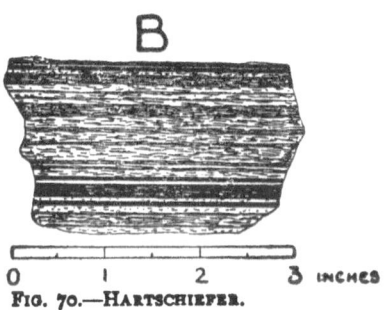

FIG. 70.—HARTSCHIEFER.

A. Hartschiefer, Kebnekaise, Swedish Lappland. After Quensel. (See text, p. 288.)

B. Hartschiefer, Mt. Bourrée, Prince Charles Foreland, Spitsbergen. After Tyrrell. (See text, p. 288.)

Along zones of intense and rapid movement, the development of frictional heat may be so great that local fritting and fusion of mineral particles may occur. The rock may thus be locally vitrified, forming black streaks (pseudo-tachylyte) in the mass, and structureless isotropic material in thin section.

This glassy matter ramifies in irregular veins through the rock, and contains angular fragments of the mylonised groundmass. Rocks of this character, which often become chert-like or felsite-like in texture, are variously called *ultramylonite, flinty crush rock*, and where occurring in gneissose rocks, *trap-shotten gneiss*.

Associated with mylonites and ultramylonites, and apparently due to recrystallisation from them, are rocks known as *hartschiefer*, which are of compact, dense, cherty or felsitic texture, and exhibit banding of the most rigid parallelism and equality of thickness (Fig. 70). The bands differ considerably in mineral and chemical composition. Quensel's hartschiefer from Swedish Lappland have been manufactured from syenite, through the intermediate stages of mylonite-gneiss (flaser-gneiss) and mylonite-schist (mylonite)[1] The rock must have been under extreme conditions of solution, or even fusion, with almost complete recrystallisation, as a considerable amount of diffusion must have taken place in order to form the separate bands. If hartschiefer can be formed from massive igneous rocks, they should be produced with even greater facility from bedded, banded, and laminated sedimentary rocks, as has been shown to be the case in Prince Charles Foreland, where hartschiefer have been formed from banded siltstones, quartzites, and crush-breccias.[2]

---

[1] Quensel, *op. cit. supra*, pp. 103 ; 108-16.
[2] Tyrrell, *op. cit. supra*, pp. 464-5.

CHAPTER XVIII

THERMAL METAMORPHISM AND ITS PRODUCTS

THERMAL METAMORPHISM IN GENERAL—Thermal metamorphism includes pyrometamorphism, contact metamorphism optalic metamorphism, and pneumatolytic metamorphism, but these classes are not sharply marked off, and pass by insensible graduations one into the other. By *pyrometamorphism* is here meant the effects of the highest degree of heat possible without actual fusion, acting under relatively-dry conditions. *Contact metamorphism* takes place at lower temperatures, and mineral transformations are facilitated by abundance of rock moisture aided by magmatic emanations. The changes produced with positive additions of material from magmatic sources are dealt with later under the heading of pneumatolytic metamorphism (p. 324). The character and extent of the changes produced by the above kinds of metamorphism depend on a number of variable factors. The greater the size of an intrusive mass the greater, in general, the degree of metamorphism effected in the surrounding rocks. Thus, the *contact aureole* around large granite batholiths may be several miles wide. In small sills and dykes, however, the alteration may be limited to a few feet or inches immediately adjoining the contacts.

Again, the higher the initial temperature, and the slower the rate of cooling, the greater is the metamorphic effect produced. These factors are to some extent functions of the depth within the crust at which intrusion takes place. The rate of cooling is perhaps the most important. The walls of a volcanic pipe through which magma has been discharged for a long period of time are frequently most intensely altered; while lavas emitted at the surface, although at a high initial temperature, produce only a feeble baking effect upon the subjacent rocks.

It is also found, other things being equal, that granites produce a much greater metamorphic effect than other plutonic rocks. This is mainly due to the fact that acid magmas are much more highly charged with chemically-active gases and liquids than are other magmas ; and as these substances soak into the adjacent rock, metamorphic changes are thereby greatly facilitated.

Finally, the varying composition and texture of the rocks subjected to heat influence the nature and extent of the metamorphism. The effects of original composition have already been dealt with (p. 260). The stability-fields of different minerals and mineral assemblages vary widely. Some react quickly to temperature changes ; others do not react at all. An open, porous texture is more favourable to metamorphism than a close, dense, compact texture, as it permits freer diffusion of the magmatic fluids and rock moisture which play so large a part in contact metamorphism.

THERMAL METAMORPHISM OF CLAY ROCKS—Clay rocks are composed chiefly of particles of quartz and felspar, flakes of mica and chlorite, and a flour of colloidal decomposition products such as hydrous aluminium silicates, iron oxides, and silica. Colloidal clay matter has a selective affinity for potash salts, and to a less extent, for titanic acid ; hence these substances are often added to clay rocks by adsorption. The chemical constituents of clays important from the point of view of metamorphism are therefore silica, alumina, mag-

Substances present.	Minerals formed.
SiO_2	Quartz.
Al_2O_3	Corundum.
TiO_2	Rutile.
SiO_2, Al_2O_3	Andalusite, chiastolite, sillimanite.
SiO_2, $(Fe, Mg)O$	Orthopyroxenes.
SiO_2, Al_2O_3, $(K, Na)_2O$	Muscovite, alkali-felspars.
SiO_2, Al_2O_3, CaO, Na_2O	Plagioclase felspars.
SiO_2, Al_2O_3, $(Mg, Fe)O$	Cordierite, magnesia-garnet.
SiO_2, Al_2O_3, $(Fe, Mg)O$	Staurolite, iron-garnet (almandine).
SiO_2, $(Mg, Fe)O$, CaO	Diopside.
SiO_2, Al_2O_3, $(Mg, Fe)O$, $(K, Na)_2O$	Biotite.

nesia, ferrous oxide, potash, and ferric oxide, and in a smaller degree, soda, lime, and titanium dioxide. In contact metamorphism these constituents unite to form certain characteristic minerals, especially silicates of alumina, many of which are unknown in igneous rocks (see Table, p. 290).

These minerals are formed within different temperature-pressure fields from clay rocks of appropriate composition. The mineral assemblages which are produced from systems of three, four, and even five oxide components under thermal metamorphism can be represented on triangular composition diagrams.[1] The most fundamental diagram for the study of the thermal metamorphism of argillaceous rocks is that in which the corners represent respectively Al_2O_3, $(Mg,Fe)O$, SiO_2 (Fig. 71). From mixtures represented by points within the subsidiary triangles those minerals crystallise, and are in equilibrium, which are represented at the corners of the triangles. The following table shows the minerals which form by thermal metamorphism from mixtures represented by points within the triangles 1, 2, 3, etc.

Triangle.	Assemblage.
1	Quartz, cordierite, orthopyroxenes (enstatite, hypersthene).
2	Quartz, cordierite, andalusite (low temp. form), sillimanite (high temp. form).
3	Cordierite, andalusite or sillimanite, spinel.
4	Cordierite, andalusite or sillimanite, corundum
5	Andalusite or sillimanite, corundum, spinel.
6	Corundum, spinel, cordierite.
7	Cordierite, spinel, orthopyroxene.
8	Spinel, orthopyroxene, forsterite.
9	Spinel, forsterite, (periclase).

In dealing with argillaceous rocks, the assemblages 8 and 9, containing magnesia-rich minerals, need not be considered. Assemblages of composition 1 to 7, due to the thermal metamorphism of various types of argillaceous rocks, are known as *hornfels*. Many hornfels contain alkali-felspar, plagioclase, biotite, and quartz, in addition to the minerals given above;

[1] C. E. Tilley, *Geol. Mag.*, 60, 1923, pp. 101-7; 410-8. See also Grubenmann-Niggli, *Gesteinsmetamorphose*, I, 1924, pp. 376-94.

and these are due to the presence of the alkali components
K$_2$O and Na$_2$O which, if in relatively small quantity, do not
appear to disturb the equilibrium relations at all seriously.

The four (or five) component system,

CaO, (MgFe)O, Al$_2$O$_3$, SiO$_2$

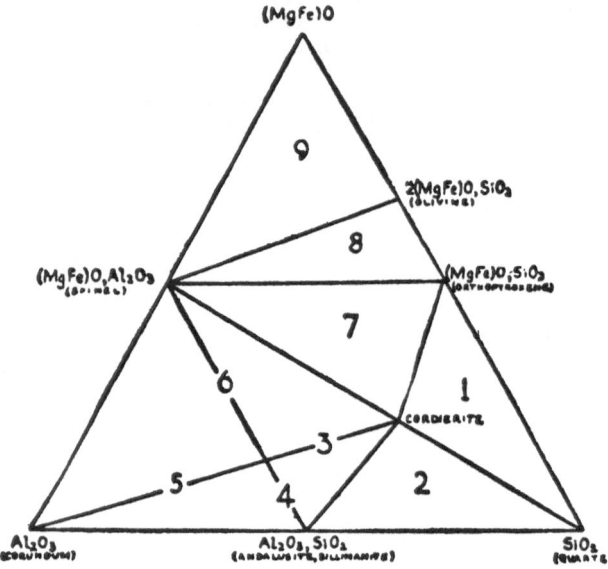

FIG. 71.—EQUILIBRIUM DIAGRAM OF THE SYSTEM (MgFe)O, Al$_2$O$_3$,
SiO$_2$, UNDER THERMAL METAMORPHISM.[1]

The corners of the triangles show the minerals developed by thermal
metamorphism from mixtures represented by points within the triangles.

can be treated in a triangular diagram if silica be regarded as
" free," and as occurring in all possible combinations
(Fig. 72).[2] Under conditions of thermal metamorphism

[1] This diagram is actually that for the system MgO, Al$_2$O$_3$, SiO$_2$;
but FeO may replace MgO up to a certain limit without bringing in any
new phases, although the shapes of the stability fields may be somewhat
altered.

[2] C. E. Tilley, *op. cit.*, p. 411 ; Grubenmann-Niggli, *op. cit.*, pp. 387-8.

there may crystallise from the various fields the mineral assemblages shown in the accompanying table:—

Triangle.	Assemblage.
1	Quartz, orthopyroxene, cordierite, anorthite.
2	Quartz, orthopyroxene, diopside, anorthite.
3	Quartz, diopside, wollastonite, grossular (garnet).
4	Quartz, diopside, grossular, anorthite.
5	Quartz, anorthite, cordierite, andalusite or sillimanite.

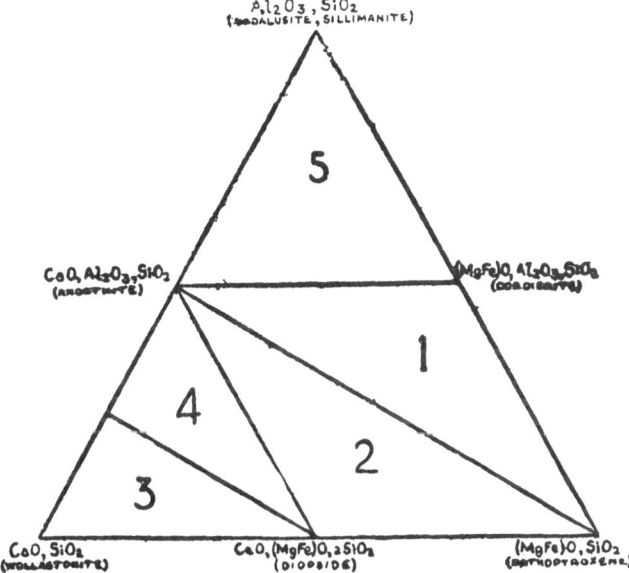

FIG. 72.—EQUILIBRIUM DIAGRAM OF THE SYSTEM CaO, (Mg, Fe)O, Al₂O₃, SiO₂, UNDER THERMAL METAMORPHISM.
(For explanation, see text, p. 292.)

With alkali components also present anorthite is modified into plagioclase of varying composition; alkali-felspar and biotite may also be formed.

CONTACT METAMORPHISM OF CLAY ROCKS—The classic example of the contact metamorphism of an argillaceous rock is that of the Steiger slate of Alsace, where it comes into contact with the great granite masses of Barr-Andlau and Hohwald.[1] The sediments are fairly uniform, and it is possible to trace definite zones of alteration as follows :—

1. Outermost zone of spotted slate. The spots are segregations of carbonaceous matter, but the mass of the rock remains practically unaltered.

2. Intermediate zone of spotted hornfels. The rocks are harder, and the cleavage has been obliterated. The spots are now incipient crystals of andalusite, and the groundmass has been recrystallised with the abundant development of mica and quartz.

3. Innermost zone of andalusite hornfels. Adjacent to the granite margin the spots disappear, and there has been thorough recrystallisation into a fine-grained granoblastic aggregate of andalusite, biotite, muscovite, and quartz.

A good example of the contact metamorphism of an already metamorphosed clay rock is provided by the pelitic schists (Dalradian) of the Glencoe region, where they enter the aureoles of the Ballachulish and Ben Nevis granites.[2] The original rock was a phyllite or fine-grained mica-schist, with porphyroblasts of biotite, chlorite, and less often garnet, in a base of muscovite, quartz, and occasionally alkali-felspar. The contact metamorphism of this aggregate has resulted in the production of andalusite, cordierite, and alkali-felspar, with the partial or complete destruction of muscovite, chlorite, garnet, and quartz. Biotite remains unaltered, or is simply recrystallised. Andalusite and orthoclase appear to have been formed in accordance with the equation—

$$H_2KAl_3(SiO_4)_3 + SiO_2 = Al_2SiO_5 + KAlSi_3O_8 + H_2O$$
(muscovite)   (quartz) (andalusite) (orthoclase)

By an analogous reaction cordierite is produced from a combination of biotite, chlorite, and garnet, with additional alumina and silica, which is probably obtained from the muscovite and quartz of the rock.

---

[1] H. Rosenbusch, Abh. 2, *Specialkarte von Elsass-Lothringen*, Strassburg, 1877.

[2] "Geology of Ben Nevis and Glencoe," *Mem. Geol. Surv. Scotland* 1916 pp. 195-201.

The contact metamorphism of argillaceous rocks containing a variable amount of lime, and thus ranging from pure clay-slate to marl, has been fully described by V. M. Goldschmidt from the Oslo (Kristiania) district of Norway,[1] where the Lower Palæozoic sediments have been metamorphosed by members of an alkaline igneous complex. From considerations based on the geological structure it is estimated that the metamorphism took place under a static pressure of 23625 lb. per square inch, or about 400 atmospheres ; while a temperature of from 1000°-1200° C. in the innermost contact zone is inferred from the presence and co-existence of certain minerals. Nevertheless, there can have been no extensive solution or diffusion of material, for bedding features and even graptolites are well preserved in some of the hornfels. Under the definite temperature and pressure conditions only certain minerals and assemblages of minerals can be formed from initial materials of a given composition. The following table shows the variation in mineral composition of the hornfels produced from rocks varying from pure shale at the one end to a marl (shale plus limestone) at the other. Quartz and orthoclase, as more or less neutral minerals, may appear in all varieties.

	Class of Hornfels.	Andalusite.	Cordierite.	Plagioclase.	Biotite.	Hypersthene.	Diopside.	Grossular.	Vesuvianite. Wollastonite.
Shale	1	x	x	x (Ab)	x				
	2	x	x	x	x				
	3		x	x	x				
	4		x	x	x	x			
	5			x	x	x			
	6			x	x	x	x		
	7			x	x		x		
	8			x			x		
	9			x			x	x	
Marl	10							x	x

Increasing amount of lime

On this basis Goldschmidt has instituted ten classes of hornfels arising from sediments of the composition-range shale to marl, which have been found to be of wide application.

[1] *Die Kontakt-metamorphose im Kristianiagebiet*, 1911, 483 pp.

PYROMETAMORPHISM OF CLAY ROCKS—Pyrometamorphism
is a term used by Brauns to denote the changes which the
schist xenoliths enclosed within the Laach trachyte of the
Eifel have suffered.[1]   The schist fragments have been converted into " sanidinites," consisting essentially of alkali-
felspars, cordierite, spinel, corundum, with sometimes biotite,
sillimanite, and almandine garnet.   These rocks are believed
to have been produced by a combination of melting and
pneumatolytic metamorphism.   The term pyrometamorphic
might be usefully extended to all products of the action of
very high magmatic temperatures, whether aided or not by
the chemical action of magmatic substances.   Pyrometamorphic effects are conterminous with, and are hardly distinguishable from, those due to assimilation and hybridisation (p. 163).

An excellent British example of pyrometamorphism is
provided by the sapphire-bearing xenoliths in certain tholeiite dykes of Mull.[2]   The metamorphism was not effected
by the enclosing magma, but in deep-seated magmatic chambers the lining of which consisted of a highly-aluminous,
shaly sediment.   The first effect was the actual fusion of
the material to a glass (buchite), containing needles of mullite
($3Al_2O_3, 2SiO_2$),[3] and a small amount of corundum (0·5 per
cent.) representing the excess of alumina over the requirements of mullite and the glass.   Subsequent chemical
action by the magma, involving the transference of lime,
ferrous iron, and magnesia, produced a coarsely crystalline
mass of anorthite, enclosing crystals of sillimanite, corundum
(sapphire), and spinel, with a residuum of unaltered glass.
This material was then shattered and fragments (the xenoliths) carried up into their present positions.   The alteration
of the glass to crystalline material is never symmetrical to
the margins of the xenoliths, showing that the action has
not taken place *in situ*.   The relations of the xenolithic

[1] *Der Kristallinen Schiefer des Laacher-See Gebietes und ihre Umwandlung zu Sanidinite*, Stuttgart, 1911.
[2] H. H. Thomas, *Quart. Journ. Geol. Soc.*, 78, pt. 3, 1922, pp. 229-60.
[3] Formerly regarded as sillimanite (Al_2O_3, SiO_2), and shown by Bowen,
Greig, and Zies, to be identical with the " sillimanite " of artificial melts
(*Journ. Wash. Acad. Sci.*, 14, 1924, p. 183 ; *Journ. Amer. Ceramic Soc.*,
7, 1924, p. 238).

minerals suggest that the temperature within the magmatic chamber at the period of metamorphism was between 1400° C. and 1250° C.

THERMAL METAMORPHISM OF LIMESTONES—Carbonate rocks are particularly susceptible to metamorphism because of the solubility of their minerals, the ease with which they recrystallise under conditions of augmented temperature and pressure, and the chemical reactivity of lime and magnesia. This instability is greatly increased by the presence of siliceous and aluminous impurities, as pointed out by Harker (p 261).

When calcium carbonate is heated with free access of air it is dissociated with the formation of quicklime—

$$CaCO_3 \rightarrow CaO + CO_2$$

But when the calcite is heated under pressure, as, for example, by an igneous intrusion, dissociation is hindered, the carbon dioxide is retained, and the mineral merely recrystallises as a granoblastic aggregate, forming *crystalline limestone* or *marble*. The pure white marble used as statuary and for tombstones comes chiefly from Carrara (Italy), and is due to the metamorphism of a Triassic limestone. The Greek statuary marbles came from Pentelikon, near Athens (Cretaceous), and from the island of Marmora (Eocene) near Constantinople.

Many limestones, however, contain admixtures of siliceous and aluminous (clay) impurities. The basic principle in the thermal metamorphism of impure limestones is that the lime tends to unite with the foreign material, with the elimination of carbon dioxide. If there is sufficient silica present the whole of the rock may be transformed into lime silicate. Experimental work shows that these reactions can take place in dry mixtures at about 500° C. The chief minerals formed under these conditions are lime-garnet, vesuvianite, anorthite, wollastonite, diopside, malacolite, tremolite, zoisite, and epidote. When silica is present as the sole impurity, wollastonite is formed under the appropriate conditions of temperature and pressure (Fig. 65, p. 264), according to the equation—

$$\underset{\text{(calcite)}}{CaCO_3} + \underset{\text{(silica)}}{SiO_2} \rightarrow \underset{\text{(wollastonite)}}{CaSiO_3} + CO_2$$

The formation of anorthite requires the presence of aluminous matter, and probably a rather high temperature. The reaction may be as follows :—

$$CaCO_3 + Al_2O_3,2SiO_2,2H_2O = CaAl_2Si_2O_8 + 2H_2O + CO_2$$
(calcite)   (kaolin)   (anorthite)

Anorthite is found along with vesuvianite and lime-garnet in the limestone blocks ejected from Vesuvius.

Many limestones are initially more or less magnesian without being dolomitic (p. 228). On contact metamorphism the magnesia reacts with silica and alumina to form various amphiboles and pyroxenes (tremolite, diopside, malacolite). Often a finely-granulose aggregate of calc-magnesian-silicates of this character, along with calcite, is produced, forming a *calc-silicate-hornfels*. The purer varieties of the Ballachulish limestone, containing only about 2 per cent. of MgO, and a little silica, have been metamorphosed with the production of tremolite, where they enter the aureole of the Ben Nevis granite.[1] Regarding the magnesia present as in the form of dolomite, the reaction can be represented by the following equation :—

$$3CaMg(CO_3)_2 + 4SiO_2 = CaO,3MgO,4SiO_2 + 2CaCO_3 + 4CO_2$$
(dolomite)   (silica)   (tremolite)   (calcite)

The greater part of the calcite present does not take part in this reaction, and is simply recrystallised.

The contact metamorphism of dolomitic rocks under a sufficiently low pressure is governed by the principle of the greater reactivity of the magnesian than of the calcic component. Thus the heating of pure dolomite under a comparatively low pressure results in the dissociation of the magnesian carbonate, with the crystallisation of calcite :—

$$CaMg(CO_3)_2 \rightarrow CaCO_3 + MgO + CO_2$$
(dolomite)   (calcite)   (periclase)

This destruction of dolomite and reconstitution of calcite was called *dedolomitisation* by Teall.[2] Periclase is readily hydrated to form brucite (MgO,H_2O), and the final product

[1] " Geology of Ben Nevis and Glencoe," *Mem. Geol. Surv. Scotland*, 1916, p. 190.
[2] *Geol. Mag.*, 1903, p. 513.

of the change is then *brucite-marble*, varieties of which are called *predazzite* and *pencatite* (from Tyrolean localities).

Dedolomitisation is facilitated by the presence of siliceous or argillaceous impurities, and a great variety of new minerals is produced on contact metamorphism. With a small amount of silica, forsterite (magnesian olivine) is formed :—

$$2CaMg(CO_3)_2 + SiO_2 = 2CaCO_3 + Mg_2SiO_4 + 2CO_2$$
$$\text{(dolomite)} \quad \text{(silica)} \quad \text{(calcite)} \quad \text{(forsterite)}$$

giving rise to *forsterite-marble*. By the hydration of forsterite, or of other varieties of magnesian olivine with some of the iron olivine molecule present, serpentine of various colours is formed, resulting in the production of beautifully-coloured *serpentine-marbles* or *ophicalcite*. One of the most notable ophicalcites of this country is the green-streaked marble of Glen Tilt, Perthshire.

With a somewhat larger quantity of silica, diopside is formed instead of forsterite :—

$$CaMg(CO_3)_2 + 2SiO_2 = CaMg(SiO_3)_2 + 2CO_2$$
$$\text{(dolomite)} \quad \text{(silica)} \quad \text{(diopside)}$$

An intermediate amount of silica results in the formation of tremolite according to the equation given above (p. 298). A fine tremolite marble is associated with the ophicalcite of Glen Tilt.

With alumina as an impurity in the original dolomite-rock, spinel may be formed :—

$$CaMg(CO_3)_2 + Al_2O_3 = MgAl_2O_4 + CaCO_3 + CO_2$$
$$\text{(dolomite)} \quad \text{(alumina)} \quad \text{(spinel)} \quad \text{(calcite)}$$

and with silica in addition, forsterite may be formed along with spinel :—

$$3CaMg(CO_3)_2 + Al_2O_3 + SiO_2 = Mg_2SiO_4 + MgAl_2O_4 + 3CaCO_3 + 3CO_2$$
$$\text{(dolomite)} \quad \text{(alumina)} \text{(silica)} \quad \text{(forsterite)} \quad \text{(spinel)} \quad \text{(calcite)}$$

The production of alkali-felspar (especially microcline), associated with malacolite or aluminous diopside, has been noted in many localities,[1] and is probably due to the thermal

---

[1] Glencoe : *Geology of Ben Nevis and Glencoe*, 1916, pp. 192-4 ; Cornwall : " Geology of the Country around Bodmin and St. Austell," *Mem. Geol. Surv.*, 1909, p. 100 ; South Australia : C. E. Tilley, *Geol. Mag.*, 57, 1920, p. 493.

metamorphism of dolomite with micaceous and siliceous impurities. Tilley suggests the following reaction :—

$$4MgCO_3 + 3CaCO_3 + H_2KAl_3(SiO_4)_3 + 7SiO_2 = \begin{cases} 3CaMg(SiO_3)_2 \\ MgAl_2SiO_6 \end{cases}$$

$$\text{(Ca and Mg carbonates)} \qquad \text{(sericite)} \qquad \text{(silica)} \quad \text{(aluminous diopside)}$$

$$+ KAlSi_3O_8 + 7CO_2 + H_2O$$

$$\text{(microcline)}$$

THERMAL METAMORPHISM OF ARENACEOUS ROCKS—Pure quartz and felspar sandstones, affected by a sufficient degree of heat, are merely recrystallised into granoblastic aggregates of those minerals, with the complete obliteration of their clastic characters. A peculiar vitreous lustre is often imparted to the rock, which is known as *quartzite*. Other quartzites are formed by silica cementation, and to distinguish metamorphic quartzites, the term *quartz-hornfels* might be introduced. If argillaceous, calcareous, or magnesian impurities are present in quantity as cement or otherwise, they are transformed on contact metamorphism into the minerals appropriate to their collective composition, as detailed above, but the large excess of quartz remains neutral and simply recrystallises. The Silurian flagstones and graywackes present good examples of the thermal metamorphism of impure arenaceous rocks where they enter the aureoles of the Galloway granites.[1] Coarsely crystalline sillimanite- and cordierite-bearing biotite-hornfels, sometimes with marked gneissose structures, have been produced. There is an extraordinary development of large crystals of sillimanite, along with andalusite and cordierite, where the Moine schists and gneisses enter the aureole of the Ross of Mull granite.[2] The sillimanite appears to be due to the instability of muscovite under conditions of intensive thermal metamorphism. The reaction has been represented by E. B. Bailey as proceeding according to the equation—[3]

$$H_2KAl_3(SiO_4)_3 + SiO_2 = Al_2SiO_5 + KAlSi_3O_8 + H_2O$$

$$\text{(muscovite)} \qquad \text{(quartz)} \quad \text{(sillimanite} \quad \text{(orthoclase)}$$

$$\text{or andalusite)}$$

[1] Miss Gardiner, *Quart. Journ. Geol. Soc.*, 46, 1890, pp. 569-80 ; Sir J. J. H. Teall in *Mem. Geol. Surv. Scotland*, "Silurian Rocks of Scotland," 1899, pp. 644-7.

[2] T. O. Bosworth, *Quart. Journ. Geol. Soc.*, 66, 1910, p. 376.

[3] "Pre-Tertiary Geology of Mull, L. Aline, and Oban," *Mem. Geol. Surv. Scotland*, 1925, p. 52.

The pyrometamorphism of siliceous rocks is illustrated in artificial materials by the formation of tridymite and cristobalite in the silica bricks used in steel furnaces.[1]   Dr. H. H. Thomas has described natural examples from Mull,[2] where xenoliths of sandstone have been enveloped in very hot tholeiite magma.   The felspathic constituents have been melted with the formation of an interstitial liquid which has attacked the quartz grains, and has become enriched in dissolved silica.   From this melt tridymite (SiO_2) crystals have separated, and have attached themselves as fringes to the quartz grains.   On cooling the melt has solidified as a glass, while the tridymite has reverted to quartz, retaining, however, its characteristic shape.

THERMAL METAMORPHISM OF BASIC LAVAS AND TUFFS— Acid igneous rocks show contact metamorphic effects similar in kind and degree to those of the arenaceous rocks ; but basic igneous rocks and tuffs are much more susceptible and exhibit remarkable changes.   The effects of the contact metamorphism of the basalt lavas of Skye by plutonic masses of gabbro and granophyre are first shown in the amygdales,[3] where zeolites have been converted into felspars, along with epidote, zoisite, and actinolite.   In the body of the rock the augite has been converted into fibrous hornblende, and chlorite into biotite, at some distance from the contacts ; but close to the plutonic masses the basalt has been totally recrystallised with the formation of a granoblastic pyroxene-felspar hornfels.

OPTALIC METAMORPHISM—Milch [4] has designated the in-durating, baking, burning, and fritting effects of lava-flows and small dykes on neighbouring rocks as *caustic metamor-phism*.   The term " caustic," however, is now used mainly in the sense of corrosion, eating-away ; and the word *optalic*, derived from the Greek *optaleos* = baked (as bricks), conveys the sense better, and has therefore been adopted for this

---

[1] A. Scott, *Trans. Ceramic Soc.*, 17, 1917-18, pp. 137-52 ; 459-85. Also see " Min. Res. Gt. Britain," *Mem. Geol. Surv.*, xvi, 1920, p. 60.
[2] *Quart. Journ. Geol. Soc.*, 78, 1922, pp. 239-40.
[3] A. Harker, " The Tertiary Igneous Rocks of Skye," *Mem. Geol. Surv. Scotland*, 1904, pp. 50-3.
[4] *Die Umwandlung der Gesteine. Grundzüge der Geologie*, I, 1922, p. 288.

kind of metamorphism. These optalic effects are produced by evanescent hot contacts at which the heat is rapidly dissipated. The elimination of water and other volatile constituents, the bleaching of carbonaceous rocks by the burning-off of the carbon, the reddening of iron-bearing rocks by the oxidation of the iron, induration, peripheral fusion of grains (fritting) ; in short, analogous kinds of alteration to those produced artificially in brick and coarse earthenware manufacture, are the most notable effects of this phase of metamorphism. Argillaceous rocks are often indurated with the production of an excessively hard material called *hornstone*, *lydian-stone*, or *porcellanite*. Some *honestones* or *novaculite* are due to this action on siliceous clays and shales. The coking of coal seams by igneous intrusions, and the columnar structures induced both in coals and in some sandstones, are also to be regarded as effects of optalic metamorphism.

CHAPTER XIX

DYNAMOTHERMAL METAMORPHISM AND ITS PRODUCTS

DYNAMOTHERMAL METAMORPHISM IN GENERAL—Dynamo-thermal metamorphism is due to the co-operation of directed pressure and heat. The heat element facilitates recrystallisa-tion ; but the stress element not only promotes recrystallisa-tion, but is powerful in deforming the rocks, and producing new structures. The new parallel textures and structures are usually orientated perpendicular to the direction of greatest stress, and parallel to that of minimum stress. The deforma-tion may be regarded as due to the interaction of three pro-cesses, which may be designated as *clastic*, *plastic*, and *blastic*.[1] The clastic process is instrumental in actual frac-ture, rupture, and rolling-out of the minerals as detailed in Chapter XVII. Plastic deformation occurs when, under powerful confining pressure, rocks and minerals are made to stretch and flow by movement along cleavage and gliding planes without actual rupture (fractureless deformation), perfect cohesion being maintained during the movement. In a long series of experiments, F. D. Adams and his colla-borators have shown that under sufficient pressure rocks and minerals can be deformed without fracture, and that such susceptible materials as marble can be forced to flow into, and fill, open spaces without loss of cohesion. A cylindrical column of marble 35 mm. long and 17 mm. in diameter, was compressed into a barrel-shaped mass 17·3 mm. long and 28·8 mm. in diameter.[2] The deformed marble was found to have acquired a perfect foliation, the crystals of calcite having been flattened out so that, in thin section, they appeared as

---

[1] Grubenmann-Niggli, *Die Gesteinsmetamorphose*, I, 1924, p. 235.
[2] Adams and Coker, *Amer. Journ. Sci.*, 29, 1910, p. 465.

ribbons eight to ten times as long as they were broad.   There
was no trace of granulation ; and under the conditions of
experiment the deformation could not have been assisted by
solution effects.   Under suitable conditions of pressure and
temperature even hard and brittle minerals such as quartz
can be pressed out into ribbons.[1]

Blastic deformation is that effected by recrystallisation
processes under the operation of Riecke's principle (p. 266),
in such a way that previously-existing minerals are elongated
perpendicular to the direction of greatest pressure, and new
minerals, whose cleavage and gliding characters are well
adapted to stress conditions, grow in the same plane.   In
other words, the development of stress minerals (p. 266) is
favoured.   Leith and Mead have shown that there is a
decided convergence in chemical composition towards the
typical schist-making minerals, even in the metamorphism
of rocks originally widely divergent in composition from
mixtures of micas, amphiboles, etc. ; which takes place
principally by the elimination of substances present in
excess of the requirements of these minerals.[2]

Different minerals are differently susceptible to the operation of the above-described processes.   Under the same
conditions some are deformed rupturally, while others are
elongated by flow and recrystallisation.   It is most probable
that in ordinary dynamothermal metamorphism all three
processes are simultaneously at work, with the ultimate
production of more or less completely recrystallised rocks,
the partial or complete obliteration of old textures and structures, and the imposition of new ones.   Confirmation of the
essential contemporaneity of movement and recrystallisation
is to be gained from the phenomena of " snowball " garnets
(Fig. 73), which show a spiral arrangement of inclusions of
quartz, biotite, etc., indicating that the garnets have been
rolled along by differential movement of the matrix of the
rock, whilst still in active growth.[3]   In some cases, however,
as Sander has shown, recrystallisation is completely separated
from mechanical deformation, and follows it after an interval

[1] G. W. Tyrrell, *Trans. Roy. Soc. Edin.*, 53, 1924, p. 464.
[2] *Metamorphic Geology*, 1915, p. 292.
[3] Sir J. S. Flett in " The Geology of Ben Wyvis, Carn Chuinneag," etc.,
Mem. Geol. Surv. Scotland, 1912, p. 111.

of time. He believes that the new structures merely follow, accentuate, and pseudomorph, old parallel arrangements of the mineral constituents, such as original bedding or mylonisation schistosity.[1] Confirmation of late recrystallisation is supplied by the tendency in some rocks for porphyroblasts, such as chloritoid and staurolite, to grow athwart the directions of schistosity, indicating their entire independence of, and posteriority to, the movements which gave rise to the directional features.

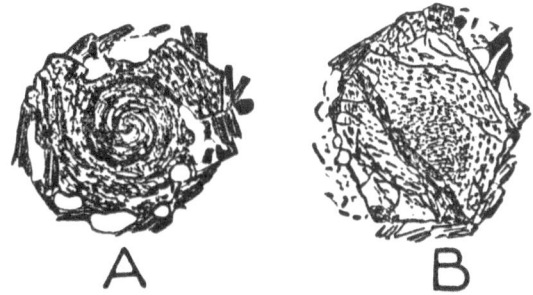

FIG. 73.—DYNAMOTHERMAL METAMORPHISM. SNOWBALL
GARNETS.

A. From sheared garnetiferous biotite-hornfels, Salachie Burn, Glen Calvie, Ross-shire. After *Mem. Geol. Surv. Scotland*, " The Geology of Ben Wyvis, Carn Chuinneag, Inchbae," etc., 1912.
B. From garnetiferous mica-schist, Allt Caillich, Glen Lyon, Perthshire. This garnet is altering marginally to chlorite, and also along a diagonal fissure. (See text, p. 304.)
Magnification about 40 diameters.

DYNAMOTHERMAL METAMORPHISM OF ARGILLACEOUS ROCKS—It has been shown that even at the slate stage of the dynamic metamorphism of clay rocks there is already a considerable amount of recrystallisation. When the dynamic factor is assisted by heat the degree of recrystallisation increases rapidly. The numbers and sizes of the mica flakes are augmented, and the rock passes through the stage of *phyllite* (a still cleavable rock with visible mica flakes) into that of *mica schist*. Any excess of alumina is met by

[1] Grubenmann-Niggli, *Die Gesteinsmetamorphose*, I, 1924 pp. 234-5.
20

the formation of kyanite, the stress member of the Al_2O_3, SiO_2 assemblage of minerals. Alumina may also enter, along with $(Mg,Fe)O$, into the common pink garnet, almandine, the ferro-magnesian element being supplied by the destruction of chlorite in the original rock. Thus is formed the very common *garnetiferous mica schist* type. Any lime present may go to form zoisite, epidote, or, perhaps, hornblende ; but in schists formed from argillaceous rocks these minerals are usually subordinate in amount. They appear much more abundantly when the original rock is of marly composition, and certain types of amphibolite are thus formed. With ferruginous impurities in the original clay chloritoid or ottrelite may be formed at an early stage of metamorphism, and staurolite at a later stage. From original carbonaceous shales graphitic mica schists may be derived. At very high temperatures sillimanite may appear in pelitic schists and gneisses, and felspars may be produced at the expense of the micas, according to the reaction given on page 300.

With an increasing degree of metamorphism the grain size is enlarged, and the rocks become definitely schistose. Ultimately, at the highest stage of metamorphism, in which felspars are developed, they assume a gneissose structure. Collectively, the range of rock types produced by dynamothermal metamorphism from clay rocks may be referred to as *pelitic schists and pelitic gneisses*. The dominant mineralogical feature of these rocks is the abundance of mica, both biotite and muscovite ; but the latter may occasionally be replaced by its soda analogue paragonite. Quartz is practically always present ; and numerous other minerals, felspar, garnet, chlorite, chloritoid, staurolite, kyanite, or hornblende, may be present in subordinate amount, giving rise to minor varieties of mica-schist and mica-gneiss. Innumerable examples of mica-schists in great variety have been described from the Dalradian and Moine formations of the Scottish Highlands,[1] and from the Mona Complex of Anglesey.[2] The Moine Gneiss contains pelitic schists and gneisses in addition to its dominant psammitic types. These rocks are muscovite-biotite-schists or gneisses, with lenticles

---

[1] See *Mem. Geol. Surv. Scotland* on Highland areas.
[2] E. Greenly, " Geology of Anglesey," *Mem. Geol. Surv.*, 1923.

of quartz and felspar, which are often garnetiferous, but seldom contain kyanite or staurolite. Their composition indicates that they were originally typical argillaceous rocks.[1] The development of garnetiferous muscovite-biotite-schist by the shearing of contact-metamorphosed Moine rocks around the Carn Chuinneag (Ross-shire) granite, has been described.[2] The hornfelses, which retain the original sedimentary lamination, consist of a granoblastic aggregate of quartz, alkali-felspar, garnet, and biotite, with occasional andalusite or sillimanite, and represent fine-grained sandy shales. A later dynamothermal metamorphism has involved hornfels and granite alike, producing a perfect schistose structure and certain mineralogical changes in the hornfels, of which the most important is the destruction of felspar with the formation · of muscovite. In this change some potash is set free which helps to transform andalusite and sillimanite also into white mica. It may also combine with (Mg,Fe)O and Al_2O_3, to form biotite. This ferro-magnesian material is derived from the garnets and iron oxides present, which therefore tend to diminish in amount. The remaining garnet is regenerated and forms fewer but larger crystals than in the hornfels. It is in this rock that the " snowball " garnets figured (p. 305) have been found. The identity of the biotite-hornfels with the pelitic Moine schist has been established by chemical analysis.

THE DYNAMOTHERMAL METAMORPHISM OF QUARTZO-FELSPATHIC ROCKS—The quartzo-felspathic rock types include the acid igneous rocks with their porphyries and lavas, the sandstones, quartzites, and quartz-conglomerates. These are stubborn and resistant to alteration, and under a given intensity of metamorphism, may reach only the cataclastic stage of metamorphism, whilst associated argillaceous, calcareous, or basic rocks, may be more or less thoroughly recrystallised. They show in the greatest perfection the cataclastic and flaser structures described in Chapter XVII. The crushing-down of a rock to a mass of fine particles in the early stages of metamorphism, greatly facilitates the

[1] Sir J. S. Flett in " The Geology of the Lower Findhorn and Lower Strath Nairn," *Mem. Geol. Surv. Scotland*, 1923, pp. 53-7.
[2] Sir J. S. Flett in " The Geology of Ben Wyvis, Carn Chuinneag," etc., *Mem. Geol. Surv. Scotland*, 1912, pp. 107-12.

subsequent recrystallisation, as a very much larger aggregate surface is exposed to the action of interstitial solutions.

Pure quartz sandstones and quartzites naturally show little mineralogical change. They are merely recrystallised and rendered more or less schistose, forming quartz-schist or schistose quartzite. Parallel structures are often barely noticeable ; mosaic quartzites with orientated wisps of sericite derived from felspathic impurities, and occasionally a few lines of heavy minerals (magnetite and zircon) which mark the original bedding, make up much of the Perthshire quartzite formations. The term *quartz-schist* may be used for rocks somewhat richer in sericite, which have been produced by the dynamothermal metamorphism of moderately felspathic sandstones or quartzites.

The metamorphism of impure sandstones with a considerable amount of argillaceous or calcareous cement, or ferromagnesian impurities, results in the production of schistose grit, albite-schist, etc., rocks which are common along the southern margin of the Scottish Highlands.[1] In the schistose grits cataclastic structures are still frequent ; but the albite schists and gneisses are thoroughly recrystallised. The mineralogical constitution of these rocks is quartz ; white mica derived from alkali-felspar (p. 309) ; chlorite and biotite derived from the ferro-magnesian impurities ; albite, due to the break-up of plagioclase felspar in the manner more fully described in the next section ; epidote, zoisite, or even a little hornblende, originating from the lime of the plagioclase reinforced by any other calcareous matter which may have been present. The albite may occur as small water-clear grains intermingled with the quartz matrix, or it may form porphyroblasts which are usually crowded with inclusions.

The dynamothermal metamorphism of granites proceeds in a very similar manner to that of coarse felspathic sandstone. The first stage is the production of flaser granite. The felspars are the most resistant minerals, and remain as porphyroclasts when the rest of the rock has been crushed down into a granular aggregate. The *augen* or *eyed* structure which is thus produced is much more pronounced when

[1] E. H. Cunningham-Craig, " Metamorphism in the Loch Lomond District," *Q.J.G.S.*, 60, 1904, pp. 10-29.

the original rock was a porphyritic granite or a granite-porphyry. In the succeeding grade of metamorphism the recrystallisation of the fine-textured paste to a mixture of quartz and white mica is completed; while the porphyro-clasts may or may not be reconstituted. The final product is a coarse quartzo-felspathic gneiss, which, in view of its igneous origin, is called *orthogneiss* or *granite-gneiss*.

The characteristic mineralogical change in the dynamo-metamorphism of quartzo-felspathic rocks is the breaking-down of alkali-felspars with the production of white mica and quartz. This change, which is called *sericitisation*, begins at an early stage in metamorphism, and is reversible, especi-ally with more intense thermal action. Clarke gives the following equation as the most probable for the reaction :—[1]

$$\underset{\text{(orthoclase)}}{3KAlSi_3O_8} + \underset{\text{(water)}}{H_2O} = \underset{\text{(sericite)}}{KH_2Al_3Si_3O_{12}} + \underset{\substack{\text{(potassium} \\ \text{silicate)}}}{K_2SiO_3} + \underset{\text{(quartz)}}{5SiO_2}$$

Albite may, perhaps, be reacted upon in like manner to pro-duce paragonite (the soda mica), and sodium silicate. Clarke suggests that the liberated potassium silicate may help to develop muscovite from albite and plagioclase felspar. In any case solutions of potassium and sodium silicate thus rendered available may aid in the felspathisation and injec-tion-metamorphism of neighbouring rocks (p. 327).

Fine-grained acid igneous rocks, such as quartz-porphyries, felsites, or rhyolites, may form closely-felted sericite-schists on dynamothermal metamorphism.

The Lewisian Gneiss of the North-west of Scotland is mainly a primary igneous gneiss, in which the banding and gneissose structures are due to the flow of a heterogeneous magma. The granitic and other gneisses have been much modified along lines and zones of later shearing, with the production of a distinct foliation, and the development of white mica, epidote, and zoisite, from the felspars of the granitic types.[2]

The central Alpine granite-gneisses (*protogine*) are batho-lithic masses of granite which have been metamorphosed by the Alpine orogenetic movements. The micas and felspars

[1] *Data of Geochemistry*, 5th ed., 1924, p. 604.
[2] " The Geological Structure of the North-west Highlands of Scotland," *Mem. Geol. Surv. Scotland*, 1907, Chaps. III and IV.

are often aligned tangentially to the pressure, and there has been a considerable development of white mica, epidote, and zoisite, from the felspars. These rocks have been cited by Weinschenk as examples of *piezocrystallisation*,[1] by which is meant the crystallisation of a viscous constrained magma under the influence of directed pressure. The above-mentioned secondary minerals are thus regarded by Weinschenk as of direct magmatic origin.

DYNAMOTHERMAL METAMORPHISM OF BASIC IGNEOUS ROCKS AND TUFFS—An important group of metamorphic rocks, including the chlorite-, talc-, and hornblende-schists, amphibolites, and hornblende-gneisses, arise by the dynamothermal metamorphism of diorites, gabbros, dolerites, basalts, and ultrabasic rocks, or tuffs of like composition. The original minerals of these rocks are calcic plagioclase, augite, olivine, and iron ores, including both ilmenite and magnetite. From this assemblage, at a moderately intense degree of metamorphism, are formed albite, zoisite, epidote, hornblende, sphene, talc, anthophyllite, and garnet. At a lower grade of metamorphism chlorite tends to be formed in place of the amphiboles.

The anorthite molecule of plagioclase is unstable under dynamic metamorphism, and breaks up into zoisite, epidote, prehnite, etc., with the liberation of the albite. Under low-grade metamorphism the products of the alteration merely pseudomorph the felspars as a very dense microgranular mixture of albite, zoisite (or epidote), with variable amounts of prehnite, sericite, chlorite, actinolite, and garnet. This substance, which in pre-microscope days was thought to be a definite mineral species, is known as saussurite, and the alteration is called *saussuritisation*.[2]

At an early stage of metamorphism any ferro-magnesian mineral may break down into chlorite. Chloritisation, indeed, is often due to intensive weathering, and characterises such altered rocks as diabase and propylite. The alteration of pyroxenes to chlorite frequently gives rise to calcite and quartz as by-products. Chlorite also appears as the product of the alteration of garnet. In the next stage of

[1] *C.R. VIII, Congr. Géol. Internat.*, Paris, 1900, tom. 2, p. 326.
[2] Sir J. S. Flett and J. B. Hill, " Geology of the Lizard and Meneage," *Mem. Geol. Surv.*, 1912, p. 89.

metamorphism hornblende is produced directly from pyroxene, and epidote or zoisite and quartz may develop as by-products of the change.   This process is known as *uralitisation ;* the secondary hornblende pseudomorphous after augite being called uralite.   By the same mode of alteration diopside may yield tremolite, and soda-pyroxenes the sodic aluminous amphibole glaucophane.   Dr. C. E. Tilley states that hornblende may arise by chemical interaction between chlorite and calcite, or between chlorite and epidote, both reactions requiring the addition of silica.[1]

Under dynamothermal metamorphism olivine breaks down into tremolite, anthophyllite, or talc.   Leucoxene, a dense aggregate of minute granules of sphene, arises from the alteration of ilmenite ; and under more intense metamorphism large crystals of sphene are formed from ilmenite or other titaniferous minerals.

Many of these mineralogical changes appear to take place at the first onset of pressure before new textures are imposed. Rocks thus altered while still retaining their original textures may be called *meta-gabbro, meta-dolerite,* and *meta-basalt* respectively.   The term *epidiorite* has been applied to doleritic or basaltic rocks in which the pyroxenes have been uralitised, so that the mineralogical composition approaches that of diorite.   At the next stage of metamorphism the directed pressure breaks down the original structures and textures, and flaser-gabbros, gabbro-schists, etc., are formed.[2] Sir J. S. Flett defines a flaser-gabbro as a gabbro which exhibits signs of crushing and recrystallisation under pressure, but has neither completely lost its igneous structure, nor has assumed the characters of a schist.

Hornblende-schists with a well-marked linear foliation are the final products of the extreme dynamothermal metamorphism of basic rocks.   Quartz, plagioclase, albite, biotite, epidote, zoisite, and rutile, may occur as subordinate constituents of these rocks.   The term *amphibolite* is used for a hornblendic rock in which foliation gives place to granoblastic texture.   Garnet is a frequent and often abundant accessory in either type, giving rise to garnetiferous hornblende-schist and garnetiferous amphibolite.   The garnet

[1] *Q.J.G.S.*, 79, pt. 2, 1923, p. 197.
[2] *Geology of the Lizard and Meneage*, 1912, pp. 87-91.

may be formed by a reaction between original olivine and the anorthite molecule of plagioclase, leaving albite as a by-product.   Excellent examples of all these types are provided by the sills of epidiorite, hornblende-schist, and amphibolite, which are interbedded with the Dalradian series of the Scottish Highlands (Ben Vrackie, etc.).[1]   Numerous hornblende-schists and amphibolites occur within the Lewisian of the North-west of Scotland, and are due to the dynamothermal metamorphism of the basic members of the original igneous complex.   In the famous Scourie dyke, a member of a later series of intrusions intersecting the gneiss but still of Lewisian age, Sir J. J. H. Teall traced a complete transition from a central ophitic dolerite, through massive hornblende rock, to well-foliated hornblende-schists on the margins, where shearing movements had been localised.[2]

[1] " Geology of Blair Atholl, Pitlochry, and Aberfeldy," *Mem. Geol. Surv. Scotland*, 1905, pp. 52 ; 77-83 ; Pls. II and IV.
[2] *Q.J.G.S.*, 41, 1885, p. 137.

CHAPTER XX

PLUTONIC METAMORPHISM AND ITS PRODUCTS

PLUTONIC METAMORPHISM IN GENERAL—By *plutonic meta-morphism* is meant the changes which are produced in rocks by great heat and uniform pressure. These changes neces-sarily take place in the kata-zone of Grubenmann, at depths wherein directed pressure becomes less and less pronounced, and finally becomes a practically negligible factor in meta-morphism. The high temperature which is also a char-acteristic of this depth-zone is maintained by the natural increase of heat in depth due to the temperature gradient, and by magmatic heat. Geological study of deeply-eroded regions of the earth's crust shows that the lower levels are almost everywhere penetrated by igneous intrusions on a far greater scale than the upper parts of the crust; and as these lower levels in general consist of the oldest rocks, it is the Archæan basement (the *Grundgebirge* of German and Scandinavian geologists) which most often shows the effects of plutonic metamorphism. As the regional intrusion of magma, especially magma of granitic composition, is a char-acter of great depth, the problems of plutonic metamor-phism are more or less closely connected with the problems of the soaking of rocks in magmatic emanations, their wholesale injection by igneous veins and sheets, and their final melting and incorporation within the magma. These are the phenomena of injection-metamorphism, granitisa-tion, etc., which are treated at greater length in the succeed-ing chapter. In this chapter it will be desirable to disregard these effects, and to consider the phenomena caused entirely by the co-operation of high temperature and great hydro-static pressure, with little or no addition of magmatic material.

As the influence of directed pressure is at most feeble in

this type of metamorphism, oriented parallel structures are, in general, unimportant, and give place to even-grained, granulose, directionless structures. Radical recrystallisation without marked directional tendencies thus occurs under the conditions of plutonic metamorphism, and such rocks as granulite, eclogite, and granulitic gneisses, are the typical products. The formation of anti-stress minerals of small specific volume and high density, will obviously be favoured by these conditions. The rock types produced are cordierite-, sillimanite-, and garnet-gneisses due to the metamorphism of rocks ranging from argillaceous to arenaceous composition ; pyroxene-gneiss, eclogite, garnet-amphibolite, and jadeite, derived from basic igneous rocks ; granulite and leptite from quartzo-felspathic rocks ; crystalline limestones and quartzites. Certain pyroxenic igneous rocks, as, for example, the charnockite series, are believed to have acquired their distinctive characters under the conditions of plutonic metamorphism. Many of the rock types have a close resemblance to the more extreme products of thermal metamorphism.

GRANULITE, LEPTYNITE, LEPTITE—This group of rocks covers a wide range of composition, but, in general, it may be regarded as derived from plutonic metamorphism of quartzo-felspathic rocks. The typical *granulites*, from the famous granulite region of Saxony,[1] are light-coloured rocks consisting of xenoblastic felspar and quartz grains, frequently with abundant pyroxene and garnet, and occasionally sillimanite, kyanite, green spinel, etc., in small quantity. The typical texture is that of a granulose gneiss, as the rocks often have a parallel banded structure due to the streaky elongation of quartz grains (Fig. 68B), and to the alternation of bands of different mineral composition (leptynite). The chemical composition of granulites is that of granite or felspathic sandstone. There is often an excess of alumina over alkalies and lime which results in the formation of kyanite, sillimanite, almandine, and hercynite (green spinel). The Saxon granulites occur as a complex of lenticular form 31 miles long by 11 broad, together with bands of pyroxene-granulite, biotite-gneiss, cordierite-gneiss, garnet-rock, and amphibolite. The whole complex appears to have been

[1] J. Lehmann, *Untersuchungen über die Entstehung der altkristallinen Schiefergesteine im Sächsischen Granulitgebirge*, Bonn, 1884.

derived from igneous rocks, granites and gabbros, with subordinate sedimentary intercalations. Granulites occur in
many of the great Archæan basements, as in Peninsular
India, West Africa, Scandinavia, Canada, and Antarctica.
In the North-west of Scotland many types of the Lewisian
Gneiss should be referred to as granulite. Thus rocks consisting of quartz-felspar mosaics with accessory hornblende,
biotite, iron ore, sphene, and rutile, which might be called
hornblende-granulite, occur near Loch Inver.[1]

The Moine Gneiss formation, which covers an enormous
area in the Central and Northern Highlands of Scotland,
provides good examples of paragneisses which have, in the
main, been developed by plutonic metamorphism with the
co-operation of a certain amount of stress, from a series of
grits, felspathic sandstones, and shales. The predominant
psammitic variety of the Moine gneiss is largely a granoblastic aggregate of quartz and alkali-felspar, through which
muscovite and biotite are scattered as separated flakes with
parallel orientation.[2] There are many accessory minerals,
zoisite, epidote, zircon, magnetite, ilmenite, sphene, apatite,
and rutile; and these are sometimes disposed in bands or
lines which follow the original bedding.

Leptite is a term now extensively used by Scandinavian
geologists to indicate fine-grained granulose metamorphic
rocks, which are composed chiefly of quartz and felspars, with
subordinate mafic minerals, including biotite, hornblende,
and occasionally garnet. They have a wide range of composition from almost pure quartzite to amphibolite, and undoubtedly represent a great variety of original rock types.
Thus, in Finland, Eskola distinguishes blastoporphyritic
leptites which have been derived from porphyritic acid
igneous rocks, and even-grained leptites probably of sedimentary or tuffaceous origin. A cordierite-leptite has the
composition of an argillaceous sediment.[3] Sederholm regards

---

[1] " Geol. Structure of the North-west Highlands of Scotland," *Mem.
Geol. Surv. Scotland*, 1907, pp. 63-5, Pl. XLV.
[2] G. Barrow, " On the Moine Gneisses of the East Central Highlands,"
Q.J.G.S., 60, 1904, pp. 400-49; Sir J. S. Flett in *The Geology of Ben
Wyvis, Carn Chuinneag*, etc., 1912, Chap. III; and in *Geology of the
Lower Findhorn and Lower Strath Nairn*, 1923, pp. 53-7.
[3] " Petrology of the Orijärvi Region," *Bull. Comm. Géol. Finlande*,
No. 40, 1914, pp 131 *et seq.*

leptite as merely the "sack-name" for a group to which
fine-grained granulose metamorphic rocks of doubtful origin
are relegated.[1] Leptitic rocks clearly derived from granite-
porphyry and aplite occur in Benguella (Portuguese West
Africa), where they are intrusive into an Archæan basement,
and have been immersed, as have the Finnish examples, in
immense granite batholiths.[2] In both regions, as also in
Peninsular India, rocks of this character are associated with
scapolite- and cordierite-bearing granulites, and with members of the charnockite series.

PYROXENE-GNEISS, PYROXENE-GRANULITE, AND THE CHAR-
NOCKITE SERIES—Pyroxene-bearing rocks are abundant under
conditions of plutonic metamorphism. Associated with the
common types of the Saxon granulites there are dark varieties
which are rich in pyroxenes (augite and hypersthene) and
plagioclase. Alkali-felspar may sometimes occur, garnet is
often an important accessory, and there are quartz-rich,
quartz-poor, and quartz-free varieties. Most of these rocks
have a typical granulose structure (pyroxene-granulite) ;
others, however, have a more or less marked gneissose structure (pyroxene-gneiss). The pyroxene-granulites of Saxony
seem to have been basic igneous rocks which have acquired
their present mineral composition and texture by slow recrystallisation under conditions of great uniform pressure
and heat. The gneissose structure of some varieties may
have been impressed by load metamorphism.

Pyroxene-gneisses and granulites have been described from
many parts of the world, but almost exclusively from
Archæan basement complexes. Thus from the Lewisian
Gneiss of the North-west of Scotland, garnetiferous hyper-
sthene-augite-plagioclase-granulites have been described, in
which, as in the Saxon rocks, and in the charnockite series,
all the minerals are xenoblastic and of extraordinary freshness. Some varieties of pyroxene-granulites and gneisses in
this locality contain quartz.[3] They are believed to be all
of plutonic igneous origin.

[1] "On Migmatite and Associated Pre-Cambrian Rocks of South-
west Finland," *Bull. Comm. Géol. Finlande*, No. 58, 1923, pp. 2-4.
[2] G. W. Tyrrell, "Contribution to the Petrology of Benguella," *Trans.
Roy. Soc. Edin.*, 51, pt. 3, 1916. pp. 543-6.
[3] *Geological Structure of the North-west Highlands of Scotland*, 1907,
pp. 50-6.

The charnockite series of India and other localities, although usually ranked as igneous rocks, may also be considered in this connection. Charnockite itself is a granular hypersthene granite ; the charnockite series comprises a group ranging from charnockite, through intermediate and basic rocks (quartz-norite and norite), to pyroxenites, all characterised by the abundance of pyroxenes, especially hypersthene.[1] They are unquestionably of plutonic igneous origin, and are most often of Archæan age. The evidence for metamorphic recrystallisation is found in their well-marked xenoblastic texture, like that of a coarse hornfels ; the ideal freshness of the minerals ; the occasional presence of gneissose banding ; and the appearance of myrmekitic quartz-felspar growths (p. 94) or *quartz de corrosion*. The occasional abundance of garnet without signs of concomitant dynamic metamorphism,[2] and of sporadic round quartz grains enclosed in felspars and mafic minerals alike, may also be regarded as signs of metamorphism. This constituent possibly represents silica set free during recrystallisation, like the round quartz grains of an amphibolite. Finally, the associations of charnockite rocks are with granulites, some of which contain scapolite, cordierite, and other minerals of undoubted metamorphic origin (India, West Africa, etc.).[3]

The peculiar characters of rocks of the charnockite series may be explained in two ways : they are either of primary igneous crystallisation under conditions of high temperature and great uniform pressure ; or they represent plutonic igneous rocks of the usual characters which have undergone slow recrystallisation in the solid state on being subjected to conditions of plutonic metamorphism. In this connection it may be noted that the analyses of the charnockite series listed by Washington [4] do not exhibit any marked differences from those of rocks of the normal granite to gabbro series ; and it is therefore conceivable that the charnockite series represents these rocks transformed by recrystallisation without notable chemical change.

---

[1] Sir T. H. Holland, " The Charnockite Series," *Mem. Geol. Surv. India*, 28, pt. 2, 1900, pp. 119-249.
[2] *Ibid.*, p. 196.   [3] Tyrrell, *op. cit. supra*, pp. 539-46.
[4] *Amer. Journ. Sci.*, 41, 1916, pp. 323-38.

Eclogite and Garnet Amphibolite—*Eclogite* is a coarse-grained granulose rock consisting of garnet and pyroxene. Rutile, iron ores, and apatite, may occur as subordinate constituents; and various other minerals, such as quartz, kyanite, sillimanite, felspars, bronzite, and olivine, may appear in special varieties. Hornblende may replace pyroxene to such an extent that the rock passes over to the group of garnet-amphibolite. The pyroxene is frequently the bright-green variety known as omphacite, in which the soda-pyroxene jadeite enters into solid solution with ordinary augite. Most frequently eclogites and the associated garnet-amphibolites occur in Archæan formations. Their modes of occurrence and chemical composition suggest derivation from basic igneous rocks, especially dolerite and gabbro. Under conditions of plutonic metamorphism lime-soda-plagioclase is unstable; the anorthite molecule reacts with original olivine or augite so as to form garnet; and the albite molecule enters into the pyroxene to form omphacite. Jadeite itself appears in many Alpine eclogites. Any alumina remaining after the above reactions are completed may unite with silica to form kyanite or sillimanite.

Numerous occurrences of eclogite belonging to the Lewisian are found in the Glenelg district of Sutherlandshire. They occur in thin seams and irregular masses associated with garnet-amphibolite, and interbanded with garnet-biotite-gneiss and crystalline limestone.[1]

Eclogites may also be of primary igneous origin, being due to the direct crystallisation of a basic magma under conditions of great hydrostatic pressure. Thus, for example, in the eclogites of Norway,[2] there are certain occurrences forming bands and lenses in olivine-rock, which are believed to represent late segregations from the ultrabasic magma. These eclogites as a whole are poor in alkalies and rich in chromium oxide. In the same region (Nordfjord and Møre) eclogite also appears as lenticular masses in granite-gneiss, and these rocks differ from those associated with the ultrabasic intrusions in being almost free from chromium oxide

[1] "Geology of Glenelg, Lochalsh, and South-east part of Skye," *Mem. Geol. Surv. Scotland*, 1910, pp. 32-5.
[2] P. Eskola, "On the Eclogites of Norway," *Videnskapsselsk. Skr.*, 1, *Math.-Nat. Kl. Kristiania*, No. 8, 1921, pp. 218.

and in containing relatively abundant alkalies. The lenses have been amphibolised along their margins by later movements (Fig. 74). Eskola believes that this variety of eclogite represents the dark segregations common in granites, which have been recrystallised, along with the surrounding granite-gneiss, under conditions of plutonic metamorphism.[1]

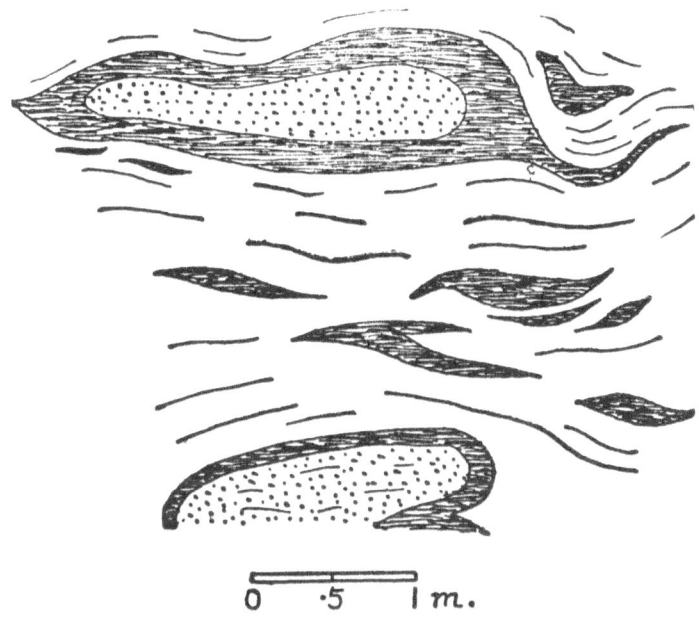

FIG. 74.—PLUTONIC METAMORPHISM. ECLOGITE.
Lenses of eclogite in granite-gneiss, Bryggen, Nordfjord, Norway. The larger masses are amphibolised along their margins, while the smaller ones are completely altered to amphibolite. After Eskola, *The Eclogites of Norway*, 1921.

Nodular lumps of eclogite-like rocks occur enclosed in the kimberlite (mica-peridotite) of the South African diamond pipes. These masses consist essentially of garnet and chrome-

[1] P. Eskola, " On the Eclogites of Norway," *Videnskapsselsk. Skr.*, 1, *Math.-Nat. Kl. Kristiania*, No. 8, 1921, pp. 63-4.

diopside, with accessory bronzite, brown mica, olivine, or kyanite. Occasionally, also, diamonds and graphite are enclosed in the nodules. These nodules have been regarded as fragments brought up from a very deep-seated eclogite formation ; [1] but according to Wagner [2] they represent fragments of segregations or schlieren of garnet-pyroxene rocks (like the *ariegite* of Lacroix [3]), which were scattered through the deep-seated peridotite zone in which the kimberlite magma was generated. This mode of origin is closely similar to that advocated by Eskola for the igneous eclogites of Norway (see above).

[1] V. M₁ Goldschmidt, " Der Gesetze der Gesteinsmetamorphose," *Videnskapsselsk. Skr.* 1, *Math. Nat. Kl. Kristiania*, 1912, No. 22, pp. 16.
[2] *The Diamond Fields of South Africa*, 1914, p. 130.
[3] *Compt. Rend.*, Paris, tom. 165, 1917, p. 381.

CHAPTER XXI

METASOMATISM AND ADDITIVE PROCESSES OF METAMORPHISM

METASOMATISM—The rôle of interstitial rock moisture and solutions of foreign derivation, as the media of mineral change in metamorphism, has been emphasised in the foregoing pages. This process of mineral or rock alteration by the agency of solutions is called *metasomatism* (change of body). Metasomatism, of course, is not confined to metamorphic rocks, but goes on, in varying degree, in all kinds of material. Lindgren [1] has recently defined metasomatism as " an essentially simultaneous molecular process of solution and deposition by which, in the presence of a fluid phase, one mineral is changed to another of differing chemical composition." A distinction may be drawn between metasomatism taking place in open spaces and in liquids or colloidal media, and those going on within capillary spaces as in metamorphism. In the former case the reactions are governed, in general, by the laws of mass action and the phase rule. Reactions in capillary spaces, however, are complicated by surface energy phenomena, and the ordinary chemical laws do not appear to apply in all cases. Further, in open spaces and in yielding media, metasomatism usually takes place with appropriate changes of volume ; but in a solid medium the replacement is not molecule for molecule, but particle for particle, or volume for volume. This view is based on good geological evidence showing that little or no change of volume has occurred in rocks which have undergone metamorphic change, and on the frequent preservation of textures and structures in altered rocks. [2]

---

[1] *Bull. Geol. Soc. America*, 36, 1925, p. 247.
[2] W. Lindgren, *Journ. Geol.*, 26, 1918, pp. 542-54.

Lindgren's conception of metasomatism applies to minerals, and is only applied to rocks when the composition of the "whole rock" is changed.[1] V. M. Goldschmidt, however, views metasomatism as the transformation of a rock with essential changes in its composition. His full definition is as follows : " Metasomatism is a process of alteration which involves enrichment of the rock by new substances brought in from the outside. Such enrichment takes place by definite chemical reactions between the original minerals and the enriching substances." [2] In effect, this definition narrows down metasomatism to alterations and replacements effected in rocks by solutions of external origin which bring in new substances ; whereas Lindgren makes the term practically synonymous with replacement by any kind of solution generated within or without the rock, and whether new substances are brought in or not. Both these conceptions have a place in the discussion of metamorphic rocks, but Goldschmidt's view is more applicable to the types of metamorphism dealt with in the present chapter. According to Goldschmidt the laws of mass action apply, on the whole, to metasomatism. In order to effect replacement the solutions must reach a definite minimum concentration in each case. This fact explains why metasomatic processes are common, but not so universal as the solutions themselves, since the solutions can only occasionally, and under special circumstances, reach the minimum concentration necessary to effect replacement.

METASOMATIC PROCESSES [3]—Metasomatic processes may be classified as follows, according to the kinds of rock they affect : (1) Metasomatism of silicate rocks and silica, illustrated by the numerous replacements and alterations which go on in metamorphic rocks, some of which are described in more detail below ; (2) Metasomatism of carbonate rocks, illustrated by the alteration of limestone to dolomite or siderite ; (3) Metasomatism of salt deposits, illustrated by the changes which take place in the highly soluble sodium

[1] If, however, metasomatism takes place in individual minerals of a rock (i.e. in a mineral aggregate) the composition of the rock must necessarily be changed, and one might as well speak of the metasomatism of the rock. The term "rock metasomatism" must be used in any case to denote the alteration of a cryptocrystalline or glassy rock.

[2] *Econ. Geol.*, 16, 1922, p. 106.   [3] *Ibid.*, p. 108 *et seq.*

and potassium salt beds ; (4) Metasomatism of sulphide rocks, illustrated by the well-known enrichment reactions in sulphidic ore deposits.

In metamorphism we are mostly concerned with metasomatic reactions in silicate and carbonate rocks. Goldschmidt classifies metasomatism in silicate rocks according to whether there has been addition of metallic compounds, or of non-metals and their compounds. In the first group of processes a distinction can be made between alkali, magnesia, lime, iron, and nickel metasomatism, according to the metallic compounds introduced. Alkali metasomatism is represented by the processes of albitisation and analcitisation, the felspathisation of schists (injection metamorphism), and the formation of biotite, muscovite, and aegirine with the aid of alkali-rich solutions. The abundant occurrence of magnesian solutions in rocks is attested by the widespread dolomitisation of limestones. Magnesian solutions are capable of dissolving silica, especially the colloidal silica which is liberated by weathering (p. 123). These silicated magnesian solutions attack quartz and acid silicates with the formation of anthophyllite and cordierite ; and under the intensified conditions of metamorphism existing within the aureoles of great granite masses, have produced cordierite-anthophyllite-gneisses, such as those of Finland.[1] Lime metasomatism is illustrated by the formation of epidote at the expense of micas, and iron metasomatism by many processes through which basic igneous rocks, slates, and limestones have been replaced by ferriferous solutions to form iron ores. Serpentine has the property of precipitating the silicate of nickel from nickeliferous solutions in the form of garnierite. The replacement of serpentine by garnierite is then an example of nickel metasomatism.

Metasomatism by the addition of non-metals and their compounds may be classified as halogen metasomatism, sulphur metasomatism, phosphorus metasomatism, silica metasomatism, and water and carbon dioxide metasomatism. Halogen metasomatism involves the addition of fluorine and chlorine (frequently with boron). It is fully dealt with later

[1] P. Eskola, " On the Petrology of the Orijärvi Region," *Bull.* 40, *Comm. Géol. Finlande*, 1914, p. 262.

under the heading of pneumatolytic metamorphism. Sulphur metasomatism is illustrated by the process of pyritisation, whereby iron pyrites is formed by the reaction of sulphide solutions with the iron in rocks and by the formation of alunite ($K_2O,3Al_2O_3,4SO_3,6H_2O$) from felspars. Phosphorus metasomatism is represented by the replacement of rocks by phosphates under a capping of guano (p. 240). The widespread phenomena of silicification, with the replacement of rocks by flint, chert, opal, and chalcedony, may be cited as examples of silica metasomatism. Water and carbon dioxide metasomatism is the chemical basis of the ordinary processes of weathering, with the formation of serpentine from olivine, chlorite from biotite and pyroxene, zeolites and carbonates from felspars, etc. Intensive alteration on these lines in intermediate and basic lavas by the action of hot carbonated waters as a sequel to vulcanism is the essential feature of the process of propylitisation (p. 127).

PNEUMATOLYTIC METAMORPHISM—Pneumatolytic metamorphism may be defined as the alteration in rocks due to the combined effects of heat and magmatic emanations largely consisting of the halogen elements, water, and compounds of boron, phosphorus, and the alkali metals. The term *pneumatolysis* (= gas action) indicates the metasomatic processes effected by these agents, mainly in the gaseous state, and at a high temperature. The chief minerals produced include muscovite, lithia-mica, fluorite, topaz, tourmaline, axinite, apatite, and scapolite. The alteration may affect the igneous rock itself (endogenous), as well as the adjacent country rock (exogenous).

The chemical elements concerned in pneumatolysis vary with the nature of the magma. With granites, the reacting substances, beside water, are fluorine, boron, compounds of the alkali metals (including lithium and beryllium), and compounds of a special suite of metals such as tin, copper, lead, zinc, tungsten, molybdenum, and uranium. With basic magmas, water, along with chlorine, and compounds of phosphorus and titanium are the chief substances concerned in pneumatolysis.

In granite pneumatolysis the minerals formed vary with the nature of the emanations and of the materials with which they react. When, along with the ubiquitous water, boron and fluorine are predominant, tourmaline (a complex silicate of aluminium and boron, with variable amounts of magnesium, iron, and alkalies) is formed; with boron alone, axinite (borosilicate of calcium and aluminium) may be produced by reaction with calcareous rocks. When fluorine is the most prominent constituent of the magmatic emanations, fluorite (calcium fluoride), and topaz (aluminium silicate with fluorine and hydroxyl) are formed. With alkalies and superheated waters, muscovite and lithia mica are produced; with beryllium, the mineral beryl (aluminosilicate of beryllium) appears; and with the appropriate constituents many other metallic compounds, such as cassiterite (dioxide of tin), are formed.

Three main types of pneumatolysis are connected with the intrusion of granitic magma : tourmalinisation, greisening, and kaolinisation. *Tourmalinisation* is due to the combined action of water, boron, and fluorine, which are concentrated in the residual liquors towards the end of the crystallisation period of the granite. They may attack the already solidified parts of the igneous mass, and the felspars are partially replaced by tourmaline, with the formation of *tourmaline-granite*. With more intense activity the felspars are completely destroyed, and the rock is then converted into an aggregate of quartz and tourmaline, which is termed *schorl-rock*. The Roche Rock of St. Austell, Cornwall, is a fine example of the complete tourmalinisation of a granite.[1] When this process affects the country rock already metamorphosed by the granite, there is formed tourmaline-hornfels, tourmaline-schist, or tourmaline-slate (cornubianite), according to the texture of the rock. The tourmaline here accompanies andalusite and cordierite.[2]

Greisening is a process of metasomatic alteration due to the action of superheated steam and fluorine. In granite the felspars are attacked and converted into white mica, which is often lithium-bearing. Hence there result aggregates of muscovite and quartz, which are called *greisen*. Albite seems to resist pneumatolysis of this type, and is retained in the greisen when the potash-felspar has been

[1] Sir J. S. Flett in " The Geology of St. Austell," *Mem. Geol. Surv.*, 1909, pp. 65-8.
[2] *Ibid.*, p. 64.

entirely destroyed. Topaz is often an important constituent of greisen, and may become so abundant as to constitute a *topaz-rock*. The adjacent country rocks are extensively muscovitised by the process of greisening, and topaz and fluorite are also introduced.

Kaolinisation is due chiefly to superheated steam aided by a little fluorine and boron. The felspars af the granite are attacked, with the formation of the mineral kaolinite $(Al_2O_3, 2SiO_2, 2H_2O)$ the main constituent of china clay.

Pneumatolytic effects are much less common in connection with the intrusion of basic rocks than with granites, and when they occur, they are found to be due to the action of chlorine, phosphorus, titanium, and their compounds, along with the ever-present water. Veins of apatite (chlor-apatite, $Ca_3(PO_4)_2, CaCl_2$), and of rutile (TiO_2) are here analogous to the fluorite, tourmaline, and tinstone veins associated with granitic intrusions. By the introduction of chlorine into the felspar molecule the mineral scapolite is produced. Scapolite forms a mix-crystal series analogous to the plagioclase felspars, with marialite as the soda-rich end member corresponding to albite, and meionite as the lime-rich end member corresponding to anorthite. In the process of *scapolitisation* the plagioclase first becomes riddled with small enclosures containing a saturated solution of common salt with floating crystals ; later on these enclosures are absorbed into the molecular structure of the crystal, with the formation of perfectly fresh scapolite.[1]

INJECTION METAMORPHISM AND AUTOMETAMORPHISM—In the foregoing section was discussed the metasomatic action of the highly volatile constituents of magmas. We now go on to describe the metamorphic effects due to the injection of the partial magmas, not volatile but still very mobile, which become available towards the end of the process of crystallisation. These residual magmas are rich in water, alkalies, alumina, and silica ; that is, in the components of alkali-felspars. Their injection into rocks, or imbibition by rocks, lead, therefore, to processes of alkali metasomatism, and especially felspathisation. Before the residual liquids leave the parent rock they may attack the previously-formed

[1] J W. Judd, *Min. Mag.*, 8, 1889, p. 186.

minerals, producing effects which are discussed under the heading of autometamorphism.

Two processes may be distinguished in exogenous metamorphism of this type: the imbibition of alkaline solutions from the magma, and the injection of residual liquors of pegmatitic or aplitic nature between the planes of fissility of the invaded rock. It has been shown (p. 309) that some metamorphic reactions have simple silicates of potassium and sodium as by-products; and the passage of these solutions into rocks, such as slates, in which there is an excess of available alumina, leads to the process of felspathisation. This process is one of soaking-in, or imbibition, of alkaline solutions of magmatic origin. On the other hand, the injection process is a more forcible one in which magmatic material is bodily squirted along the bedding, cleavage, or schistosity planes of the affected rocks. There are, of course, all transitions between the injection of acid and alkaline magmatic residua, and of undifferentiated magma. The final result is the production of lit-par-lit gneiss, veined gneiss, etc., as detailed in the succeeding section.

A typical case of imbibition and injection metamorphism is that described by Goldschmidt from the Stavanger district of Norway.[1] Here phyllites of Cambro-Silurian age, consisting essentially of quartz, muscovite, and chlorite, come into contact with masses of granite and quartz-diorite. In the outermost zone of the contact aureole, from 1 to 4 kms. from the igneous margin, the phyllites show small rounded spots of garnet. Closer in, the grain-size of the rock increases, the muscovite is partially replaced by biotite, and an enrichment in albite begins. This mineral first appears in the groundmass, and later on as porphyroblastic crystals. Hornblende and clinozoisite are formed at this stage in addition to garnet. Still closer to the contact microperthite begins to appear, and along with the albite, forms streaks, folia, and augen. The innermost contact zone consists of an injection-gneiss in which the magmatic residues have been forcibly injected into the parting planes of the rock. Chemically, the increasing metamorphism is characterised by

[1] " Die Injektionmetamorphose in Stavanger-Gebiete," *Videnskaps-selsk Skr.*, I, *Math.-Nat Kl., Kristiania*, 1921, No. 10, 142 pp.

additions of silica and soda, possibly also lime, and the loss of water. The newly-formed albite is believed to be due to the combination of sodium silicate solutions (" water-glass ") derived from the igneous rock, with the excess alumina of the phyllites. In the change from phyllite to albite-schist there is a considerable increase of volume, and this probably accounts for the peculiar contorted or ptygmatic folding (p. 334) of the metamorphosed rocks.

The chemical activity of granitic emanations is shown by the formation of certain gneisses in the Pre-Cambrian of Mozambique,[2] where granitic magma has been injected into a series of ancient sediments. The argillaceous facies became granitised with the formation of biotite-gneiss, whilst the calcareous and dolomitic facies formed hornblendic and garnetiferous gneisses by interaction with the granite.

A peculiar type of metamorphism occurs in zones a few feet wide on the margins of small basic intrusions into slate or shale, especially when the intrusions are the albite-rich types belonging to the spilite kindred. The argillaceous rock is altered into a dense, horny, rock of conchoidal frac- ture, which consists of a fine-grained mixture of quartz and albite, with small variable amounts of actinolite, rutile, chlorite, or epidote, the slaty structure being completely obliterated. These rocks are called *adinole* when thoroughly transformed ; *adinole slate* when the slaty materials and structure are partially retained. Chemical analyses of the unaltered slates, and of the adinole resulting from their alteration, show that there has been a considerable trans- fusion of silica and soda from the igneous to the metamor- phosed rock. Rocks of this character are abundant in Cornwall, where albite-diabases and " greenstones " have come into contact with the killas or slate of that region.[1]

With the exogenous injection-metamorphism due to the expulsion of residual magmatic material into the surround- ing rocks can be linked up numerous endogenous effects pro- duced by the same agents. The processes of albitisation, analcitisation, and probably serpentinisation, belong here,

[1] A. Holmes, *Q.J.G.S.*, 74, 1919, pp. 31-98.
[2] H. Dewey, *Trans. Roy. Geol. Soc. Cornwall*, 15, 1915, p. 71 ; C. Reid and J. S. Flett, " Geology of the Land's End District," *Mem. Geol. Surv.*, 1907, p. 26. See also other Cornish memoirs.

as well as the formation of reaction-structures such as myrmekite, coronas, and kelyphitic borders (p. 93). Phenomena such as these pass insensibly into those occurring in the normal reaction cycle between early-crystallised minerals and the magmatic fluid in which they are immersed (p. 76). The alteration of an igneous rock by its own residual liquors has been called *autometamorphism* by Sargent.[1] The endogenous pneumatolysis or autopneumatolysis (Lacroix) described in the preceding section must be regarded as a special case of autometamorphism. All these phenomena are on the borderline between truly igneous and truly metamorphic effects, and are variously classed by petrologists with one or the other.

R. J. Colony [2] has recently drawn attention to the powerful after-effects due to adjustments of equilibrium between the highly concentrated residual liquors and the already crystallised minerals of an igneous rock. Among these effects he includes the soaking of the earlier minerals with quartz and albite, the transformation of pyroxenes into fibrous amphiboles, and the formation of micropegmatite (? myrmekite) and serpentine. W. N. Benson has advanced the view that the larger antigorite and chrysotile serpentine masses are due to the alteration of pyroxene-bearing peridotites through the agency of magmatic waters belonging to the same cycle of igneous activity as the ultrabasic rocks themselves.[3]

The albitisation of basic igneous rocks is a widespread phenomenon of autometamorphism. Bailey and Grabham have described its occurrence in the porphyritic basalt lavas of the Lower Carboniferous in Scotland.[4] The phenocrysts of labradorite have suffered most, and in them the more calcic zones have been albitised in preference to the more sodic zones, while kernels of unaltered labradorite are preserved within the albite areas. On the other hand, the more basic lavas are the less altered, indicating that the albitisation is connected with the composition of the rock, and that the source of the albite is to be found within the rock itself. The change may be ascribed to the " self-digestion " of the

[1] H. C. Sargent, *Q.J.G.S.*, 73, 1918, p. 19.
[2] *Journ. Geol.*, 31, 1923, pp. 169-78.
[3] *Amer. Journ. Sci.*, 46, 1918, pp. 693-731.
[4] *Geol. Mag.*, 1909, pp. 250-6.

lava by its own soda-rich and silica-rich residual liquid. The attack upon the labradorite phenocrysts takes place towards the end of the magmatic period. The reaction may be expressed as follows (Eskola) :—

$$CaAl_2Si_2O_8 + Na_2SiO_3 + 3SiO_2 = 2NaAlSi_3O_8 + CaO$$

$$\text{(anorthite)} \quad \text{(sodium silicate)} \quad \quad \text{(albite)} \quad \text{(lime)}$$

Lime is thereby set free; and as it appears reasonable to suppose that some of the residual liquid remains unused, the solutions may pass out into the vesicles and fissures of the lava, to form the soda- and soda-lime zeolites which are a feature of the Scottish Carboniferous basalts.

Wells [1] and Eskola [2] have recently considered the process of albitisation in connection with the genesis of rocks belonging to the spilitic kindred. Albitisation has indubitably occurred in some of these rocks; but, as Wells shows, the initial magmas must have been correspondingly soda-rich, for all members of a long and varied suite have the same richness in albite. Eskola regards the albite in spilitic rocks as due to late magmatic reaction. In albite-clinopyroxene mixtures the eutectic point must lie very close to the albite end of the series, and the pyroxene should, therefore, crystallise first from nearly all possible mixtures. But, as the ophitic relation holds in the albite-pyroxene and albite-hornblende rocks he describes, the original felspar must have crystallised first, and must have been of calcic composition. The albite must then be due to reaction between these early crystals and a soda-rich magmatic residuum, with the loss of lime, as shown by the above equation. Eskola holds that the original magma did not differ notably from ordinary basalt; but, in view of the completeness and wide range of the replacement, it is still possible to believe that the original magma must have been relatively soda-rich.

Analcitisation is a process similar to albitisation whereby the lime-soda felspars of such rocks as teschenite are partially

[1] A. K. Wells, "The Problem of the Spilites," *Geol. Mag.*, 1923, pp. 62-74.

[2] "Petrology of Eastern Fennoscandia, I, The Mineral Development of Basic Rocks in the Karelian Formations," *Fennia*, 45, 1925, pp. 78-92.

replaced by analcite.[1] The alteration takes place around the margins of crystals, and along cracks and cleavages. Concomitantly the purplish titanaugite crystals show marginal alterations to green aegirine-bearing pyroxene, olivine to biotite and chlorite, and ilmenite to biotite. All these changes imply an increase in soda, silica, and water ; and there can be no doubt but that analcitisation is also due to late magmatic reaction, and is analogous to albitisation. That the two processes may be closely related is shown by the equation—

$$\text{Analcite} + \text{Silica} = \text{Albite} + \text{Water}$$

This suggests that analcitisation is a process connected with the presence and retention of abundant water in soda-rich basic magmas, while albitisation is the corresponding process in similar magmas but relatively poor in water.

LIT-PAR-LIT GNEISS, COMPOSITE GNEISS, ANATEXIS, AND PALINGENESIS—The part played by magmatic emanations in producing injection-gneisses has been remarked in the preceding section. The intensification of this process by the injection of the main body of the granitic magma along the planes of fissility of the neighbouring rocks, leads to the production of *banded* or *lit-par-lit gneiss*, in which folia of granitic material alternate with bands more or less altered by reaction with the transfused magmatic emanations. Country rocks which are not fissile may, however, be traversed by a multitude of veins and dykes running in all directions (*veined gneiss*). With a relatively larger influx of igneous material, so that angular fragments of the shattered and altered country rock are separated and enveloped in magma, there results the *igneous breccia* of Sederholm.[2] When by more or less thorough assimilation of the country rock, and interchange of its material with the magma, a complete admixture takes place ; and the pasty mass is at the same time rendered gneissose, partly by magmatic flow and partly by the pressure at great depths, the extreme types of metamorphic rocks known as *composite gneiss* or *migmatite*

---

[1] A. Scott, " On Primary Analcite and Analcitisation," *Trans. Geol Soc. Glas.*, xvi, pt. 1, 1916, pp. 36-45.
[2] " Om granit och gneiss," *Bull. Comm. Géol. Finlande*, 23, 1907.

(Sederholm) [1] are produced. This regional granitisation, with the production of a pasty viscous mass possessing some magmatic characters, and capable of movement, has been called *anatexis* by Sederholm. [2]

Lit-par-lit injection was recognised by Darwin near Cape

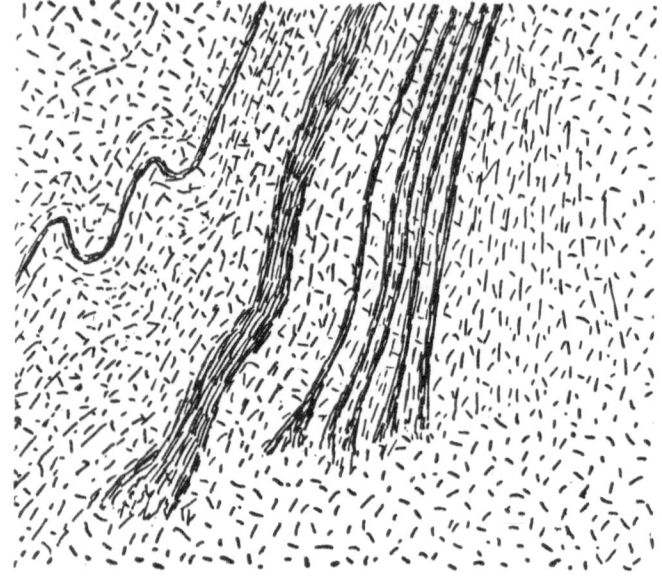

FIG. 75.—FORMATION OF COMPOSITE AND BANDED GNEISSES.

Sketch showing invasion of dark schists by granite, and the gradual fraying-out and fading of the bands, Pompton, New Jersey. After C. N. Fenner, *Journ. Geol.*, 1914. The figure represents the plan of a few square feet of glaciated rock surface.

Town, where granitic magma has insinuated itself between the folia of a schist. [3] The term *lit-par-lit*, however, was

---

[1] Gr. *migma* = a mixture. See J. J. Sederholm, "Uber die Enstehung der migmatitischen Gesteine," *Geol. Rundsch.*, 4, 1913, pp. 174-85.

[2] Gr. *ana* = up; *tektos* = melted. "On Regional Granitisation (or Anatexis)," *Cong. Géol. Internat., C.R.*, 12, 1913, pp. 319-24.

[3] *Geological Observations on Volcanic Islands*, 1844, Chap. VII.

coined by Michel-Lévy to indicate the phenomena of bed by
bed injection on the margins of certain French granites.[1]
A penetrating analysis of the phenomenon has recently
been given by C. N. Fenner in a study of the mode of
origin of Pre-Cambrian gneisses in the New Jersey highlands.[2]
The process of injection seems to have proceeded in a quiet,
gradual manner, the transfusion of igneous material into the
body of the adjacent rock, with the accompanying mutual
interactions, taking place without violent disturbance of the
attitude of the sheets or folia (Fig. 75). The preservation
of the original attitudes of the invaded rocks implies a degree

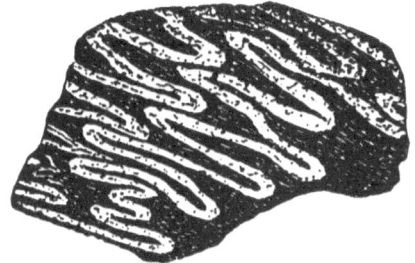

FIG. 76.—PTYGMATIC FOLDING.

Ptygmatic folding of a quartz vein in amphibolite. From a specimen
in the Lanfine Collection, Hunterian Museum, University of Glasgow.
Provenance unknown. Length of specimen, 8 inches. (See text,
p. 334.)

of viscosity within the magma that does not appear to
harmonise with the penetrative ability of the magmatic
substances. It is thought, however, that the slow advance
of the main body of magma was preceded by that of a more
dilute or even volatile portion, which exercised a preparatory
function by impregnating the country rocks, dissolving and
reacting with their minerals, and causing a general softening
or perhaps even some liquefaction. That under conditions of
deep-seated magmatic injection the invaded rock is rendered

[1] "Granite de Flamanville," *Bull. Serv. Carte Géol. France*, 5, No 36,
1893-4, pp. 317-57. Excellent Scottish examples are described by H. H.
Read, "The Geology of the Country around Golspie, Sutherlandshire,"
Geol. Surv. Scotland, 1925.
[2] *Journ. Geol.*, 22, 1914, pp 594-612 : 694-702.

pasty and viscous, is indicated by the general occurrence of areas of contorted folding, which is best shown by the more resistant layers (Fig. 76). This phenomenon has been called *ptygmatic* folding.[1]

The thoroughly mixed and reconstituted rocks known as migmatite and composite gneiss have been fully investigated by Sederholm in Finland and by Cole in Ireland. Cole's radical views on anatexis in Irish gneisses are well

FIG. 77.—ANATEXIS AND THE FORMATION OF COMPOSITE GNEISSES.
Section illustrating the intrusion of granite into folded schists with the formation of lit-par-lit and composite gneisses. After G. A. J. Cole, "On Composite Gneisses in Boylagh, West Donegal," *Trans. Roy. Irish. Acad.*, 24, 1904, p. 225. The granite is shown working its way up into a compound folded arch, and interleaving with the schists. The line SS represents the surface of the ground; below this line is the actual section; above the line is a restoration of the original structure.

summarised in his presidential address to the geological section of the British Association.[2] The banded gneisses in which granitic material alternates with biotite-schists and hornblende-schists are due, in his opinion, to the swallowing-up of mantles of basic material in huge subterranean baths of granitic magma (Fig. 77). Thus the tuffs and lavas of

[1] J. J. Sederholm, *op. cit.* (1907), p. 10; and *Neues Jahrb. f. Min.*, Beil-Bd., 36, 1913, p. 491. See H. H. Read, *Summ. Prog. Geol. Surv. Gt. Britain* for 1927, Pt. II, 1928, pp. 72-7, for a different mode of origin.
[2] *Rept. Brit. Assoc., Manchester*, 1915, pp. 407-11.

the Keewatin series are thought to have furnished the dark
basic bands in the Laurentian Gneiss of Canada, and similar
material is believed to have been worked up into the " fundamental gneiss " of Galway and Donegal, Sweden, and Finland. The late Professor J. Barrell also held the view that
regional metamorphism is due more to igneous invasion than

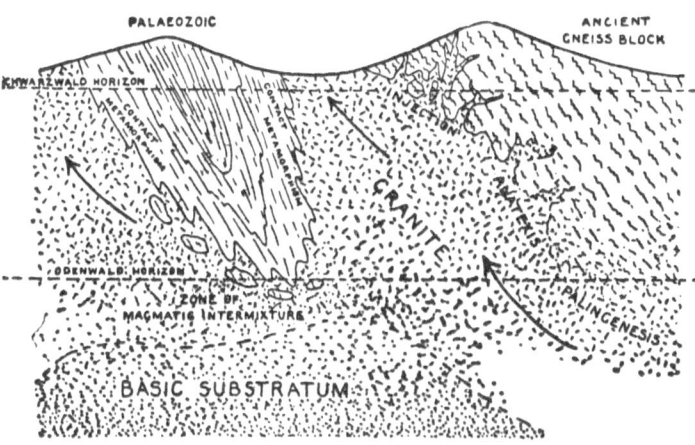

FIG. 78.—PALINGENESIS, ANATEXIS, INJECTION-, AND CONTACT-
METAMORPHISM IN SOUTH GERMANY.

The diagram shows an ancient gneiss block from the underside of which
a granite magma is being generated by fusion. An infolded Palæozoic mass is enveloped by the granite ; and with the rise of the basic
substratum beneath the geosynclinal, gabbros, diorites, and granodiorites, are formed by intermixture of the basic magma with the
granite. The upper and lower broken lines represent the horizons
at which the Schwarzwald and the Odenwald respectively intersect
the structure section. Modified from S. von Bubnoff, *Die Gliederung
der Erdrinde*, 1923.

to depth and pressure, and that recrystallisation in schists
and gneisses is largely the result of reaction with the emanations from subjacent batholiths.[1]

The above-described processes, carried to an extreme,
result in complete mixing and melting of rocks, with the

[1] " Relation of Subjacent Igneous Invasion to Regional Metamorphism," *Amer. Journ. Sci.* (5), 1, 1921, pp. 1-19; 174-86; 255 67.

production of new bodies of magma, which are capable of injection, and of passing through a cycle of igneous change. This process of regeneration of magma has been called *palingenesis* by Sederholm,[1] who ascribes to it many of the Archæan granite and granodiorite masses of Fennoscandia. Von Bubnoff [2] believes that the Hercynian granites and diorites of the Schwarzwald and the Odenwald have arisen by the regional re-fusion of the bases of ancient gneiss massifs in their vicinity (Fig. 78).

The petrogenic cycle first began on the original igneous crust of the earth, and fresh initial material has been provided with every new irruption of magma from the depths. Under the operation of external geological forces, the primary rocks have been broken down into the various types of secondary deposits. Subsequently many of the already-formed rocks, both primary and secondary, have gone through the metamorphic mill, and have been more or less completely reconstituted with new minerals and new structures. With the regeneration of magma by the regional re-melting of metamorphosed rocks in the deeper parts of the earth's crust, the great wheel of petrological change has finally turned full circle.

---

[1] Gr. *palin* = again; *genesis* = origin. Hence regeneration. In Seder-holm's original memoir (*Bull. Comm. Géol. Finlande*, No. 23, 1907), the terms anatexis and palingenesis seem to be nearly synonymous. In his subsequent publications, however, the meanings of these terms gradually approximate to those assigned to them in the text. Thus the title of one paper is "On Regional Granitisation (or Anatexis)"; and in his latest work ("On Migmatites and Associated Pre-Cambrian Rocks of South-western Finland," *Bull. Comm. Géol. Finlande*, No. 58, 1923), he describes the palingenesis or re-fusion of a conglomerate without any addition of granitic material. Palingenesis is therefore taken as the wider and more general term, indicating widespread re-fusion of rocks with or without intimate interpenetration by granite; while anatexis is assimilation and re-melting with the aid of granitic magma. Sederholm's views are further elucidated in his memoir, "The Region around the Barösundsfjord, West of Helsingfors," *Bull. Comm. Géol. Finlande*, No. 77, 1926, 143 pp.

[2] "Die Gliederung der Erdrinde," *Fortschr. d. Geol. u. Pal.*, Heft 3, 1923, pp. 44-50.

INDEX

A

Aa lava, 34.
Aberdeenshire, contaminated rocks of, 165.
Accessory minerals, 105.
Adamellite, 113.
Adams, F. D., and Coker, E. G., 303.
Additive metamorphism, 254, 320.
Adinole, 328.
Adsorption, 186.
Æolian bedding, 199.
Aerolites, 4.
Agglomerate, 15, 205.
Ailsa Craig, 114.
Albite schist and gneiss, 308.
Albitisation, 329.
Algal coal, 247.
— limestone, 238.
Alkali basalt, 128.
— granite, 112.
— peridotite, 116.
— syenite, 114.
Alkalic igneous rocks, 137, 167.
Alling, H. L., 92.
Alluvium, 211, 213.
Allogenic constituents, 189.
Allotriomorphic texture, 84-5.
Alnöite, 122.
Amorphous minerals, 222.
Amphiboles, 9.
Amphibolite, 311.
Amygdaloidal structure, 34.
Analcite basalt, 129.
— syenite, 115.
Analcitisation, 330.
Analyses, of igneous rocks, 54, 112, 117, 120, 124, 126, 131.
Anamorphism, 256.
Anatexis, 332, 334-6.
Anchi-eutectic rocks, 158.

Anderson, E. M., 25.
— Tempest, 39.
Andesite, 126.
Andrée, K., 196.
Anhedral, 83.
Ankaramite, 129.
Anorthosite, 119, 139.
— -charnockite kindred, 139, 144.
Anthracite, 247.
Anthraxylon, 245.
Anti-stress minerals, 263.
Apalhraun, 35.
Aphanitic, 83.
Aphrolith, 35.
Aplite, 113, 163.
Arber, E. A. N., 243.
Arctic kindred, 138.
Arenaceous sediments, 203, 208.
— — thermal metamorphism of, 300.
Argillaceous sediments, 203, 212.
— — (see Clay).
Arkose, 176, 209.
Arran, 40, 82, 97, 165, 176.
Arrhenius, S., 153.
Assimilation, 148, 163, 168.
Assortation of sediments, 190.
Asthenosphere, 2.
Atmosphere, 2.
Augen structure, 285, 308.
Augite, supercooling of, 58.
Augitite, 129.
Authigenic constituents, 189.
Autometamorphism, 326, 329.
Auvergne, sequence of lavas, 147.

B

Bacteria, geological work of, 226, 230, 239, 241.
Bailey, E. B., 31, 182, 237, 259, 300.